MEMOIRS
of the
American Mathematical Society

Number 950

Hypocoercivity

Cédric Villani

2000 *Mathematics Subject Classification.* Primary 35B40, 35K65, 76P05.

Library of Congress Cataloging-in-Publication Data

Villani, Cédric, 1973-
  Hypocoercivity / Cédric Villani.
    p. cm. — (Memoirs of the American Mathematical Society, ISSN 0065-9266 ; no. 950)
  "Volume 202, number 950 (fourth of 5 numbers)."
  Includes bibliographical references.
  ISBN 978-0-8218-4498-4 (alk. paper)
  1. Differential equations, Partial–Asymptotic theory. 2. Differential equations, Parabolic–Asymptotic theory. 3. Fokker-Planck equation. 4. Transport theory. I. Title.
QA377.V53 2009
515′.3533—dc22   2009029196

Memoirs of the American Mathematical Society

This journal is devoted entirely to research in pure and applied mathematics.

Subscription information. The 2009 subscription begins with volume 197 and consists of six mailings, each containing one or more numbers. Subscription prices for 2009 are US$709 list, US$567 institutional member. A late charge of 10% of the subscription price will be imposed on orders received from nonmembers after January 1 of the subscription year. Subscribers outside the United States and India must pay a postage surcharge of US$65; subscribers in India must pay a postage surcharge of US$95. Expedited delivery to destinations in North America US$57; elsewhere US$160. Each number may be ordered separately; *please specify number* when ordering an individual number. For prices and titles of recently released numbers, see the New Publications sections of the *Notices of the American Mathematical Society*.

Back number information. For back issues see the *AMS Catalog of Publications*.

Subscriptions and orders should be addressed to the American Mathematical Society, P. O. Box 845904, Boston, MA 02284-5904 USA. *All orders must be accompanied by payment.* Other correspondence should be addressed to 201 Charles Street, Providence, RI 02904-2294 USA.

Copying and reprinting. Individual readers of this publication, and nonprofit libraries acting for them, are permitted to make fair use of the material, such as to copy a chapter for use in teaching or research. Permission is granted to quote brief passages from this publication in reviews, provided the customary acknowledgment of the source is given.

Republication, systematic copying, or multiple reproduction of any material in this publication is permitted only under license from the American Mathematical Society. Requests for such permission should be addressed to the Acquisitions Department, American Mathematical Society, 201 Charles Street, Providence, Rhode Island 02904-2294 USA. Requests can also be made by e-mail to reprint-permission@ams.org.

Memoirs of the American Mathematical Society (ISSN 0065-9266) is published bimonthly (each volume consisting usually of more than one number) by the American Mathematical Society at 201 Charles Street, Providence, RI 02904-2294 USA. Periodicals postage paid at Providence, RI. Postmaster: Send address changes to Memoirs, American Mathematical Society, 201 Charles Street, Providence, RI 02904-2294 USA.

© 2009 by the American Mathematical Society. All rights reserved.
Copyright of individual articles may revert to the public domain 28 years after publication. Contact the AMS for copyright status of individual articles.
This publication is indexed in *Science Citation Index*®, *SciSearch*®, *Research Alert*®, *CompuMath Citation Index*®, *Current Contents*®/*Physical, Chemical & Earth Sciences*.
Printed in the United States of America.

∞ The paper used in this book is acid-free and falls within the guidelines established to ensure permanence and durability.
Visit the AMS home page at http://www.ams.org/

10 9 8 7 6 5 4 3 2 1   14 13 12 11 10 09

Contents

Introduction		1
Part I. $L = A^*A + B$		5
1.	Notation	7
2.	Operators $L = A^*A + B$	9
3.	Coercivity and hypocoercivity	13
4.	Basic theorem	17
5.	Generalization	23
6.	Hypocoercivity in entropic sense	28
7.	Application: the kinetic Fokker–Planck equation	40
8.	The method of multipliers	46
9.	Further applications and open problems	50
Part II. The auxiliary operator method		55
10.	Assumptions	57
11.	Main theorem	58
12.	Simplified theorem and applications	61
13.	Discussion and open problems	63
Part III. Fully nonlinear equations		65
14.	Main abstract theorem	67
15.	Proof of the Main Theorem	70
16.	Compressible Navier–Stokes system	80
17.	Weakly self-consistent Vlasov–Fokker–Planck equation	83
18.	Boltzmann equation	87
Appendices		103
A.19.	Some criteria for Poincaré inequalities	105
A.20.	Well-posedness for the Fokker–Planck equation	109
A.21.	Some methods for global hypoellipticity	113
A.22.	Local positivity estimates	127
A.23.	Toolbox	130
Bibliography		139

Abstract

This memoir attempts at a systematic study of convergence to stationary state for certain classes of degenerate diffusive equations, taking the general form $\frac{\partial f}{\partial t} + Lf = 0$. The question is whether and how one can overcome the degeneracy by exploiting commutators.

In Part I, the focus is on a class of operators taking the abstract form $L = A^*A + B$ in a Hilbert space. A general Hilbertian result is proven, which can be considered as a "spectral" counterpart of Hörmander's regularity theorem. Then I discuss an "entropic" version of this result, which leads to more general statements but needs more structure. The main example of application is the linear Fokker–Planck equation; other examples are discussed.

In Part II, a different method is discussed, based on the introduction of an auxiliary operator which has good commutation and non-commutation properties with L. Some recent results are reinterpreted in this formalism.

In Part III, a third method is discussed, applying to nonlinear equations with very little structure. This one is the most general but needs a lot of smoothness, and does not in general achieve the exponential convergence. Applications to various models of fluid mechanics, in particular the Boltzmann equation, are discussed. My recent results with Desvillettes about the convergence to equilibrium for the Boltzmann equation are extended and simplified in this way.

The unity of the three parts comes from the method: in all cases, the convergence to equilibrium is obtained by a carefully designed Lyapunov functional, or family of Lyapunov functionals. Many open problems and possible directions for future research are discussed throughout the text.

In a long Appendix, I introduce some methods for the study of global hypoellipticity, focusing on the kinetic Fokker–Planck equation once again.

Received by the editor on October 30, 2006; and in final form on October 6, 2008.
Article electronically published on July 22, 2009; S 0065-9266(09)00567-5.
2000 *Mathematics Subject Classification*. Primary 35B40; 35K65; 76P05.
Key words and phrases. Convergence to equilibrium; hypoellipticity; hypocoercivity; Fokker–Planck equation; Boltzmann equation.

©2009 American Mathematical Society

Introduction

In many fields of applied mathematics, one is led to study dissipative evolution equations involving (i) a degenerate dissipative operator, and (ii) a conservative operator presenting certain symmetry properties, such that the combination of both operators implies convergence to a uniquely determined equilibrium state. Typically, the dissipative part is not coercive, in the sense that it does not admit a spectral gap; instead, it may possess a huge kernel, which is *not* stable under the action of the conservative part. This situation is very similar to problems encountered in the theory of hypoellipticity, in which the object of study is not convergence to equilibrium, but regularity. By analogy, I shall use the word **hypocoercivity**, suggested to me by Thierry Gallay, to describe this phenomenon. This vocable will be used somewhat loosely in general, and in a more precise sense when occasion arises.

Once the existence and uniqueness of a steady state has been established (for instance by direct computation, or via an abstract theorem such as Perron–Frobenius), there are plenty of soft tools to prove *convergence* to this steady state. It is much more tricky and much more instructive to find estimates about *rates of convergence*, and this is the question which will be addressed here.

Both hypoellipticity and hypocoercivity often occur together in the study of linear diffusion generators satisfying Hörmander's bracket condition. It is for such equations that theorems of exponentially fast convergence to equilibrium were first established via probabilistic tools [41, 45, 46, 53], taking their roots in the Meyn–Tweedie theory of the asymptotic behavior of Markov chains. Some of these studies were motivated by the study of finite-dimensional approximations of randomly forced two-dimensional Navier-Stokes equations [18, 40, 41]; since then, the theory has been developed to the extent that it can deal with truly infinite-dimensional systems [30]. In all these works, exponential convergence is established, but there are no quantitative estimates of the rate. Moreover, these methods usually try to capture information about path behavior, which may be useful in a probabilistic perspective, but is more than what we need.

Analytical approaches can be expected to provide more precise results; they have been considered in at least three (quite different, and complementary) settings:

- For **nonlinear equations** possessing a distinguished **Lyapunov functional** (entropy, typically), robust methods, based on functional inequalities, time-derivative estimates and interpolation, have been developed to establish convergence estimates in $O(t^{-\infty})$, i.e. faster than any inverse power of time. These methods have been applied to the linear (!) Fokker–Planck equation [14], the Boltzmann equation [16], and some variants arising in the context of kinetic theory [9, 22]. So far, they rely crucially on strong regularity a priori estimates.

- For **linear hypoelliptic equations** enjoying some structural properties, more specific methods have been developed to prove (ideally) exponential convergence to equilibrium with explicit bounds on the rate. Up to now, this approach has been mainly developed by Hérau and Nier [34], Eckmann and Hairer [20], Helffer and Nier [32], for second-order differential operators in Hörmander's form (a sum of squares of derivations, plus a derivation). It uses pseudo-differential operators, and a bit of functional calculus; it can be seen as an extension of Kohn's celebrated method for the study of hypoellipticity of Hörmander operators. In fact, the above-mentioned works establish hypoellipticity at the same time as hypocoercivity, by considering functional spaces with polynomial weights in both Fourier space and physical space. After a delicate spectral analysis, they localize the spectrum inside a cusp-like region of the complex plane, and then deduce the exponential convergence to equilibrium. Again, in some sense these methods capture more than needed, since they provide information on the whole spectrum.

- Finally, Yan Guo recently developed a new method [28], which he later pushed forward with Strain [29, 52], to get rates of convergence for nonlinear kinetic equations in a close-to-equilibrium regime. Although the method is linear in essence, it is based on robust functional inequalities such as interpolation or Poincaré inequalities; so it is in some sense intermediate between the two previously described lines of research.

The goal of this memoir is to start a systematic study of hypocoercivity in its own right. The **basic problem considered here** consists in identifying general structures in which the interplay between a "conservative" part and a "degenerate dissipative" part lead to convergence to equilibrium.

With respect to the above-mentioned works, the novelty of the approach explored here resides in its abstract nature and its simplicity. In particular, I wish to convey the following two messages:

1. Hypocoercivity is related to, but distinct from hypoellipticity, and in many situations can be established quantitatively independently of regularity issues, or after regularity issues have been settled.

2. There are some general and simple techniques, based on very elementary but powerful algebraic tricks, by which one can often reduce a mysterious hypocoercive situation to a much more standard coercive one.

This memoir is divided into three parts.

Part I focuses on the particular case of operators which (as in [20, 32, 34]) can be written in "Hörmander form" $A^*A + B$, where A and B are possibly unbounded operators on a *given Hilbert space*. These results have been applied to several models, such as the kinetic Fokker–Planck equation, the linearized Landau–Lifschitz–Gilbert–Maxwell system in micromagnetism [10], and a model problem for the stability of the Oseen vortices [24].

Part II, by far the shortest, remains at a linear level, but considers operators which cannot necessarily be written in the form $A^*A + B$, at least for "tractable" operators A and B. In this part I shall give an abstract version of a powerful hypocoercivity theorem recently established by Mouhot and Neumann [43], explain why we cannot be content with this theorem, and give some suggestions for research in this direction.

In Part III I shall consider fully nonlinear equations, in a *scale of Sobolev-type spaces*, in presence of a "good" Lyapunov functional. In this setting I shall obtain results that can apply to a variety of nonlinear models, *conditionally to smoothness bounds*. In particular I shall simplify the proof of the main theorem in [**16**].

Though these three settings are quite different, and far from being unified, there is a unity in the methods that will be used: **construct a Lyapunov functional by adding carefully chosen lower-order terms to the "natural" Lyapunov functional**. This simple idea will turn out to be quite powerful.

The method will be presented in a rather systematic and abstract way. There are several motivations for this choice of presentation. First, the methods are general enough and can be applied in various contexts. Also, this presentation may be pedagogically relevant, by emphasizing the most important features of the problem. Last but not least, most of the time I really had to set the problems in abstract terms, to figure out a way of attacking it.

No attempt will be made here for a *qualitative* study of the approach to equilibrium, but I believe this is a very rich topic, that should be addressed in detail in the future. One of the main outcomes of my work with Laurent Desvillettes [**16**] was the prediction that solutions of the Boltzmann equation, while approaching equilibrium, would oscillate between "close to hydrodynamic" and "close to homogeneous" states. To some extent, this guess was in contradiction with a commonly accepted idea according to which the large-time behavior should be dominated by the hydrodynamic regime; nevertheless these oscillations have been spectacularly confirmed in numerical simulations by Francis Filbet. Further developments can be found in [**23**]; the results obtained by numerical simulations are so neat that they demand a precise explanation.

Research in the area of hypocoercivity is currently developing fast thanks to the efforts of several other researchers such as Thierry Gallay, Frédéric Hérau, Clément Mouhot, and others. I expect that further important results will soon be available thanks to their efforts, and hope that this memoir will become the starting point of a much more developed theory.

It is a pleasure to thank Clément Mouhot for a careful reading and many comments on the present manuscript.

Part I

$$L = A^*A + B$$

In this part I shall study the convergence to equilibrium for degenerate linear diffusion equations where the diffusion operator takes the abstract form $A^*A + B$, $B^* = -B$.

The main abstract theorem makes crucial use of **commutators**, in the style of Hörmander's hypoellipticity theorem. In its simplest form, it reduces the problem of convergence to equilibrium for the non-symmetric, non-coercive operator $A^*A + B$, to that of the symmetric, possibly coercive operator $A^*A + [A, B]^*[A, B]$. If the latter operator is not coercive, then one may consider iterated commutators $[[A, B], B]$, $[[[A, B], B], B]$, etc. in addition to just $[A, B]$.

One of the first main results (Theorem 24) can be informally stated as follows: Let $A = (A_1, \ldots, A_m)$, $B^* = -B$, and $L = A^*A + B$ be linear operators on a Hilbert space $\mathcal{H}$. Define iterated commutators C_j and remainders R_j ($1 \leq j \leq N_c$) by the identities $C_0 = A$, $[C_j, B] = C_{j+1} + R_{j+1}$ ($j \leq N_c$), $C_{N_c+1} = 0$. If $\sum_{j=0}^{N_c} C_j^* C_j$ is coercive, and the operators $[A, C_k]$, $[A^*, C_k]$, R_k satisfy certain bounds, then $\|e^{-tL}\|_{\mathcal{H}^1 \to \mathcal{H}^1} = O(e^{-\lambda t})$, where the "Sobolev" space $\mathcal{H}^1$ is defined by the Hilbert norm $\|h\|_{\mathcal{H}^1}^2 = \|h\|^2 + \sum \|C_j h\|^2$.

The key ingredient in the proof is the construction of an auxiliary Hilbert norm, which is equivalent to the $\mathcal{H}^1$ Hilbert norm, but has additional "mixed terms" of the form $\langle C_j h, C_{j+1} h \rangle$.

Applied to the kinetic Fokker–Planck equation, these theorems will yield results of convergence to equilibrium that are both more general and more precise than previously known estimates.

After this "abstract" L^2 framework, a "concrete" $L \log L$ framework will be considered, leading to results of convergence for very general data (say finite measures).

My reflexion on this subject started during the preparation of my Cours Peccot at the Collège de France (Paris), in June 2003, and has crucially benefited from interactions with many people. The first draft of the proof of Theorem 18 occurred to me while I was struggling to understand the results of Frédéric Hérau and Francis Nier [34] about kinetic Fokker–Planck equations. The construction of the anisotropic Sobolev norm was partly inspired by the reading of papers by Yan Guo [28] and Denis Talay [53]; although their results and techniques are quite different from the ones in the present paper, they were the first to draw my attention to the interest of introducing mixed terms such as $\nabla_v f \cdot \nabla_x f$. Denis also showed me a useful trick for getting long-time estimates on the moments of certain hypoelliptic diffusion equations, which is based on the construction of an adequate quadratic form.

Apart from the above-mentioned people, I was lucky enough to have fruitful discussions on the subject with Bernard Helffer, Laurent Desvillettes, Luc Rey-Bellet, Jean-Pierre Eckmann, Martin Hairer, Clément Mouhot, Stefano Olla, Pierre-Louis Lions, Patrick Cattiaux and Arnaud Guillin, as well as with Christian Schmeiser and Denis Serre, who both suggested a relation between my results and Kawashima's condition in the theory of hyperbolic systems of conservation laws.

1. Notation

1.1. Basic notation.
Let $\mathcal{H}$ be a separable (real or complex) Hilbert space, to be thought of as $L^2(\mu)$, where μ is some equilibrium measure; $\mathcal{H}$ is endowed with a norm $\|\cdot\|$ coming from a scalar (or Hermitian) product $\langle\cdot,\cdot\rangle$.

Let $\mathcal{V}$ be a finite-dimensional Hilbert space (say $\mathbb{R}^m$ or $\mathbb{C}^m$, depending on whether $\mathcal{H}$ is a real or complex Hilbert space). Typically, $\mathcal{V}$ will be the space of those variables on which a certain diffusion operator acts. The assumption of finite dimension covers all cases that will be considered in applications, but it is not essential.

Let $A : \mathcal{H} \to \mathcal{H} \otimes \mathcal{V} \simeq \mathcal{H}^m$ be an unbounded operator with domain $D(A)$, and let $B : \mathcal{H} \to \mathcal{H}$ be an unbounded antisymmetric operator with domain $D(B)$:

$$\forall h, h' \in D(B), \qquad \langle Bh, h' \rangle = -\langle h, Bh' \rangle.$$

I shall assume that there is a dense topological vector space $\mathcal{S}$ in $\mathcal{H}$ such that $\mathcal{S} \subset D(A) \cap D(B)$ and A (resp. B) continuously sends $\mathcal{S}$ into $\mathcal{S} \otimes \mathcal{V}$ (resp. $\mathcal{S}$); this assumption is here only to guarantee that all the computations that will be performed (involving a finite number of operations of A, A^* and B) are authorized. As a typical example, $\mathcal{S}$ would be the Schwartz space $\mathcal{S}(\mathbb{R}^N)$ of C^∞ functions $f : \mathbb{R}^N \to \mathbb{R}$ whose derivatives of arbitrary order decrease at infinity faster than all inverse polynomials; but it might be a much larger space in case of need.

If a linear operator S is given, I shall denote by $\|S\|$ its operator norm:

$$\|S\| = \sup_{h \neq 0} \frac{\|Sh\|}{\|h\|} = \sup_{\|h\|, \|h'\| \leq 1} \langle Sh, h' \rangle.$$

If there is need to emphasize that S is considered as a linear operator between two spaces $\mathcal{H}_1$ and $\mathcal{H}_2$, the symbol $\|S\|$ may be replaced by $\|S\|_{\mathcal{H}_1 \to \mathcal{H}_2}$.

The norm $\|A\|$ of an array of operators $(A_1, \ldots, A_m)$ is defined as $\sqrt{\sum_i \|A_i\|^2}$; the norm of a matrix-valued operator (A_{jk}) by $\sqrt{\sum \|A_{jk}\|^2}$; etc.

The identity operator $X \to X$, viewed as a linear mapping, will always be denoted by I, whatever its domain. Often a multiplication operator (mapping a function f to fm, where m is a fixed function) will be identified with the multiplicator m itself.

Throughout the text, the real part will be denoted by $\Re$.

1.2. Commutators.
In the sequel, commutators involving A and B will play a crucial role. Since A takes its values in $\mathcal{H} \otimes \mathcal{V}$ and B is only defined in $\mathcal{H}$, some notational convention should first be made precise, since $[A, B]$, for instance, does not a priori make sense. I shall resolve this issue by just tensorizing with the identity: $[A, B] = AB - (B \otimes I)A$ is an unbounded operator $\mathcal{H} \to \mathcal{H} \otimes \mathcal{V}$. In a more pedestrian writing, $[A, B]$ is the row of operators $([A_1, B], \ldots, [A_m, B])$. Then A^2 stands for the matrix of operators $(A_j A_k)_{j,k}$, $[A, [A, B]]$ for $([A_j, [A_k, B]])_{j,k}$, etc. One should be careful about matrix operations made on components: For instance, $[A, A^*]$ stands for $([A_j, A_k^*]_{j,k})$, which is an operator $\mathcal{H} \to \mathcal{H} \otimes \mathcal{V} \otimes \mathcal{V}$, while $[A^*, A]$ stands for $\sum_j [A_j^*, A_j]$, which is an operator $\mathcal{H} \to \mathcal{H}$. Also note that $[A, A]$ stands for the array $([A_j, A_k])_{j,k}$, and is therefore not necessary equal to 0. Whenever there is a risk of confusion, I shall make the notation more explicit.

1.3. Relative boundedness. Let S and T be two unbounded linear operators on a Hilbert space $\mathcal{H}$, and let $\alpha \geq 0$; then the operator S is said to be α-bounded relatively to T if $D(T) \subset D(S)$, and

$$\forall h \in D(S), \qquad \|Sh\| \leq \alpha \|Th\|;$$

or equivalently, $S^*S \leq \alpha T^*T$. If S is α-bounded with respect to T for some $\alpha \geq 0$, then S is said to be bounded relatively to T. This will be sometimes abbreviated into

$$S \preccurlyeq T.$$

Note that S and T need not take values in the same space. Of course, boundedness relative to I is just plain boundedness.

This notion can be generalized in an obvious way into relative boudedness with respect to a family of operators: An operator S is said to be α-bounded relatively to $T_1, \ldots, T_k$ if $\cap D(T_j) \subset D(S)$, and

$$\forall h \in D(S), \qquad \|Sh\| \leq \alpha \bigl(\|T_1 h\| + \ldots + \|T_k h\|\bigr).$$

If such an α exists, then S is said to be bounded relatively to $T_1, \ldots, T_k$, and this will naturally be abbreviated into

$$S \preccurlyeq T_1, \ldots, T_k.$$

1.4. Abstract Sobolev spaces. The study of partial differential equations often relies on Sobolev spaces, especially in a linear context. If one thinks of the Hilbert space $\mathcal{H}$ as a (weighted) L^2 space, there is a natural abstract definition of "Sobolev norm" adapted to a given abstract coercive symmetric operator $L = A^*A$: define the $\mathcal{H}^k$-Sobolev norm $\|\cdot\|_{\mathcal{H}^k}$ by

$$\|h\|_{\mathcal{H}^k}^2 := \|h\|^2 + \sum_{\ell=1}^{k} \|A^\ell h\|^2.$$

Here is a generalization: When some operators $C_0, \ldots, C_N$ are given (playing the same role as derivation operators along orthogonal directions in $\mathbb{R}^n$), one can define

$$(1.1) \qquad \|h\|_{\mathcal{H}^1}^2 := \|h\|^2 + \sum_{j=0}^{N} \|C_j h\|^2, \qquad \|h\|_{\mathcal{H}^k}^2 := \|h\|^2 + \sum_{\ell=0}^{k} \sum_{j=0}^{N} \|(C_j)^\ell h\|^2.$$

Of course, there is an associated scalar product, which will be denoted by $\langle \cdot, \cdot \rangle_{\mathcal{H}^1}$, or $\langle \cdot, \cdot \rangle_{\mathcal{H}^k}$.

1.5. Calculus in $\mathbb{R}^n$. Most of the examples discussed below take place in $\mathbb{R}^n$; then I shall use standard notation from differential calculus: ∇ stands for the gradient operator, and $\nabla \cdot$ for its adjoint in $L^2(\mathbb{R}^n)$, which is the divergence operator.

EXAMPLE 1. Let $x = (x_1, \ldots, x_n)$ and $v = (v_1, \ldots, v_n)$ stand for two variables in $\mathbb{R}^n$. Let $A = \nabla_v$, then ∇_v^2 is the usual Hessian operator with respect to the v variable, which can be identified with the matrix of second-order differential operators $(\partial^2 / \partial v_j \partial v_k)$ $(j, k \in \{1, \ldots, n\})$. Similary, if a and b are smooth scalar functions, then $[a\nabla_v, b\nabla_x]$ is the matrix of differential operators $[a\, \partial_{v_j}, b\, \partial_{x_k}]$.

The scalar product of two vectors a and b in $\mathbb{R}^n$ or $\mathbb{C}^n$ will be denoted either by $\langle a, b \rangle$ or by $a \cdot b$. The norm of a vector a in $\mathbb{R}^n$ or $\mathbb{C}^n$ will be denoted simply by $|a|$, and the Hilbert–Schmidt norm of an $n \times n$ matrix M (with real or complex entries) by $|M|$.

The usual Brownian process in $\mathbb{R}^n$ will be denoted by $(B_t)_{t \geq 0}$.

The notation H^k will stand for the usual Sobolev space in $\mathbb{R}^n$: explicitly, $\|u\|_{H^k}^2 = \sum_{j \leq k} \|\nabla^j u\|_{L^2}^2$. Sometimes I shall use subscripts to emphasize that the gradient is taken only with respect to certain variables; and sometimes I shall indicate a reference measure if it is not the Lebesgue measure. For instance, $\|u\|_{H_v^1(\mu)}^2 = \int |u|^2 \, d\mu + \int |\nabla_v u|^2 \, d\mu$.

2. Operators $L = A^*A + B$

For the moment we shall be concerned with linear operators of the form

$$(2.1) \qquad L := A^*A + B, \qquad B^* = -B,$$

to be thought as the negative of the generator of a certain semigroup $(S_t)_{t \geq 0}$ of interest: $S_t = e^{-tL}$. (Of course, up to regularity issues, any linear operator L with nonnegative symmetric part can be written in the form (2.1); but this will be interesting only if A and B are "simple enough".) In Proposition 2 below I have gathered some properties of L which can be expressed quite simply in terms of A and B.

2.1. Dirichlet form and kernel of L. Introduce

$$\mathcal{K} := \operatorname{Ker} L, \qquad \Pi := \text{orthogonal projection on } \mathcal{K}, \qquad \Pi^\perp = I - \Pi.$$

PROPOSITION 2. *With the above notation,*
(i) $\forall h \in D(A^*A) \cap D(B), \quad \Re \langle Lh, h \rangle = \|Ah\|^2$;
(ii) $\mathcal{K} = \operatorname{Ker} A \cap \operatorname{Ker} B$.

PROOF. The proof of (i) follows at once from the identities

$$\langle A^*Ah, h \rangle = \langle Ah, Ah \rangle = \|Ah\|^2, \qquad \Re \langle Bh, h \rangle = 0.$$

It is clear that $\operatorname{Ker} A \cap \operatorname{Ker} B \subset \mathcal{K}$. Conversely, if h belongs to $\mathcal{K}$, then $0 = \Re \langle Lh, h \rangle = \|Ah\|^2$, so $h \in \operatorname{Ker} A$, and then $Bh = Lh - A^*Ah = 0$. This concludes the proof of (ii). □

2.2. Nonexpansivity of the semigroup.
Now it is assumed that one can define a semigroup $(e^{-tL})_{t \geq 0}$, i.e. a mapping $(t, h) \longmapsto e^{-tL}h$, continuous as a function of both t and h, satisfying the usual rules $e^{0L} = \operatorname{Id}$, $e^{-(t+s)L} = e^{-tL} e^{-sL}$ for $t, s \geq 0$ (semigroup property), and

$$\forall h \in D(L), \qquad \frac{d}{dt}\bigg|_{t=0^+} e^{-tL} h = -Lh.$$

As an immediate consequence, for all $h \in D(A^*A) \cap D(B)$,

$$\frac{1}{2} \frac{d}{dt}\bigg|_{t=0^+} \|e^{-tL}h\|^2 = -\Re \langle Lh, h \rangle = -\|Ah\|^2 \leq 0.$$

This, together with the semigroup property, the continuity of the semigroup and the density of the domain, implies that the semigroup is nonexpansive, i.e. its operator norm at any time is bounded by 1:

$$\forall t \geq 0 \qquad \|e^{-tL}\|_{\mathcal{H} \to \mathcal{H}} \leq 1.$$

2.3. Derivations in $L^2(\mu)$.

In most examples considered later, the Hilbert space $\mathcal{H}$ takes the form $L^2(\mu_\infty)$, for some equilibrium measure $\mu_\infty(dx) = \rho_\infty(x)\,dx$ on $\mathbb{R}^n$, with density ρ_∞ with respect to Lebesgue measure; $\mathcal{V} = \mathbb{R}^m$, $A = (A_1, \ldots, A_m)$, and the A_j's and B are derivations on $\mathbb{R}^n$, i.e. there are vector fields $a_j(x)$ and $b(x)$ on $\mathbb{R}^n$ such that

$$A_j h = a_j \cdot \nabla h, \qquad Bh = b \cdot \nabla h.$$

In short, there is an $m \times n$ matrix $\sigma = \sigma(x)$ such that

$$A = \sigma \nabla.$$

The next Proposition presents some useful calculation rules in that context. It will be assumed that everything is smooth enough: for instance ρ_∞ lies in $C^2(\mathbb{R}^n)$ and it is positive everywhere; and σ, b are C^1. The notation σ^* stands for the transpose (adjoint) of σ.

PROPOSITION 3. *With the above notation and assumptions,*
(i) $B^* = -B \iff \nabla \cdot (b\rho_\infty) = 0$;
(ii) $A^* g = -\nabla \cdot (\sigma^* g) - \langle \nabla \log \rho_\infty, \sigma^* g \rangle$.

REMARK 4. As a consequence of Proposition 3(ii), the linear second-order operator $-L = \sum A_j^2 - (B + \sum c_j A_j)$ has the Hörmander form (a sum of squares of derivations, plus a derivation). The form $A^*A + B$ is however much more convenient for the purpose of the present study — just as in [**20**].

PROOF OF PROPOSITION 3. By polarization, the antisymmetry of B is equivalent to

$$\forall h \in \mathcal{H}, \qquad \langle Bh, h \rangle = 0.$$

But

$$\langle Bh, h \rangle = \int_{\mathbb{R}^n} (b \cdot \nabla h) h \, \rho_\infty = \frac{1}{2} \int_{\mathbb{R}^n} b \cdot \nabla(h^2) \rho_\infty$$

(2.2)
$$= -\frac{1}{2} \int_{\mathbb{R}^n} h^2 \nabla \cdot (b\rho_\infty).$$

If $\nabla \cdot (b\rho_\infty) = 0$, then the integral in (2.2) vanishes. If on the other hand $\nabla \cdot (b\rho_\infty)$ is not identically zero, one can find some h such that this integral is nonzero. This proves statement (i).

To prove (ii), let $g : \mathbb{R}^n \to \mathbb{R}^m$ and $h : \mathbb{R}^n \to \mathbb{R}$, then $\langle A^* g, h \rangle$ coincides with

$$\langle g, Ah \rangle = \int_{\mathbb{R}^n} g \cdot (\sigma \nabla h) \, \rho_\infty = -\int (\sigma^* g) \cdot \nabla h \, \rho_\infty = -\int_{\mathbb{R}^n} \nabla \cdot (\sigma^* g \rho_\infty) h$$

$$= -\int_{\mathbb{R}^n} \nabla \cdot (\sigma^* g) h \, \rho_\infty - \int_{\mathbb{R}^n} \sigma^* g \cdot (\nabla \log \rho_\infty) h \, \rho_\infty,$$

where the identity $\nabla \rho_\infty = (\nabla \log \rho_\infty) \rho_\infty$ was used. This proves (ii). □

The following proposition deals with the range of applicability for diffusion processes.

PROPOSITION 5. *Let* $\sigma \in C^2(\mathbb{R}^n; \mathbb{R}^{m \times n})$ *and* $\xi \in C^1(\mathbb{R}^n; \mathbb{R}^n)$, *and let* $(X_t)_{t \geq 0}$ *be a stochastic process solving the autonomous stochastic differential equation*

$$dX_t = \sqrt{2}\,\sigma(X_t)\,dB_t + \xi(X_t)\,dt,$$

where $(B_t)_{t \geq 0}$ *is a standard Brownian motion in* $\mathbb{R}^m$. *Then*

(i) The law $(\rho_t)_{t\geq 0}$ of X_t satisfies the diffusion equation

(2.3) $$\frac{\partial \rho}{\partial t} = \nabla \cdot (D\nabla \rho - \xi\rho), \qquad D := \sigma\sigma^*;$$

(ii) Assume that the equation (2.3) admits an invariant measure $\mu_\infty(dx) = \rho_\infty(x)\,dx$ (with finite or infinite mass), where ρ_∞ lies in $C^2(\mathbb{R}^n)$ and is positive everywhere. Then the new unknown $h(t,x) := \rho(t,x)/\rho_\infty(x)$ solves the diffusion equation

(2.4) $$\frac{\partial h}{\partial t} = \nabla \cdot (D\nabla h) - \left(\xi - 2D\nabla \log \rho_\infty\right) \cdot \nabla h,$$

which is of the form $\partial_t h + Lh = 0$ with $L = A^*A + B$, $B^* = -B$, if one defines

(2.5) $\quad \mathcal{H} := L^2(\mu_\infty); \quad A := \sigma\nabla; \quad B := \left(\xi - D\nabla \log \rho_\infty\right) \cdot \nabla.$

PROOF. Claim (i) is a classical consequence of Itô's formula. To prove claim (ii), write

$$\frac{\partial h}{\partial t} = \frac{1}{\rho_\infty}\nabla \cdot \left(D\rho_\infty \nabla h + Dh\nabla \rho_\infty - \xi\rho_\infty h\right)$$
$$= \nabla \cdot (D\nabla h) + 2D\nabla h \cdot \frac{\nabla \rho_\infty}{\rho_\infty} - \xi \cdot \nabla h$$
$$+ \frac{h}{\rho_\infty}\Big[\nabla \cdot (D\nabla \rho_\infty) - \nabla \cdot (\rho_\infty \xi)\Big].$$

As ρ_∞ is a stationary solution of (2.3), the last term in square brackets vanishes, which leads to (2.4). Define A and B by (2.5). Thanks to Proposition 3 (ii), it is easy to check that

$$A^*Ah = -\nabla \cdot (D\nabla h) - D\nabla \log \rho_\infty \cdot \nabla h,$$

so h indeed satisfies $\partial_t h + Lh = 0$. It only remains to check that $B^* = -B$. By Proposition 3 (i), it is sufficient to check that

$$\nabla \cdot (\rho_\infty(\xi - D\nabla \log \rho_\infty)) = -\nabla \cdot (D\nabla \rho_\infty - \xi\rho_\infty)$$

vanishes; but this follows again from the stationarity of ρ_∞. So the proof of Proposition 5 is complete. $\square$

2.4. Example: The kinetic Fokker–Planck equation. The following example will serve as an important application and model. Consider a nice (at least C^1) function $V : \mathbb{R}^n \to \mathbb{R}$, converging to $+\infty$ fast enough at infinity (say $V(x) \geq K|x|^\alpha - C$ for some positive constants K and C). For $x, v \in \mathbb{R}^n \times \mathbb{R}^n$, set

$$f_\infty(x,v) := \frac{e^{-[V(x)+\frac{|v|^2}{2}]}}{Z}, \qquad \mu(dx\,dv) = f_\infty(x,v)\,dx\,dv,$$

where Z is chosen in such a way that μ is a probability measure. Define

$$\mathcal{H} := L^2(\mu), \quad \mathcal{V} := \mathbb{R}^n_v, \quad A := \nabla_v, \quad B := v \cdot \nabla_x - \nabla V(x) \cdot \nabla_v,$$
$$L := -\Delta_v + v \cdot \nabla_v + v \cdot \nabla_x - \nabla V(x) \cdot \nabla_v.$$

The associated equation is the kinetic Fokker–Planck equation with confinement potential V, in the form

(2.6) $$\partial_t h + v \cdot \nabla_x h - \nabla V(x) \cdot \nabla_v h = \Delta_v h - v \cdot \nabla_v h.$$

Before considering convergence to equilibrium for this model, one should first solve analytical issues about regularity and well-posedness. It is shown by Helffer and Nier [**32**, Section 5.2] that (2.6) generates a C^∞ regularizing contraction semigroup in $L^2(\mu)$ as soon as V itself lies in $C^\infty(\mathbb{R}^n)$. To study this equation for a less regular potential V, it is always possible to regularize V into a smooth approximation V_ε, then perform all a priori estimates on the regularized problem, and finally pass to the limit as $\varepsilon \to 0$. The following well-posedness theorem justifies this procedure by forcing the convergence of the approximate solutions to the original solution.

THEOREM 6. *Let* $V \in C^1(\mathbb{R}^n)$, $\inf V > -\infty$, *and let*

$$E(x,v) := V(x) + \frac{|v|^2}{2}, \qquad \rho_\infty = e^{-E}, \qquad \mu(dx\,dv) = \rho_\infty(x,v)\,dv\,dx.$$

Then, for any $h_0 \in L^2(\mu)$, *equation* (2.6) *admits a unique distributional solution* $h = h(t,x,v) \in C(\mathbb{R}_+; \mathcal{D}'(\mathbb{R}^n_x \times \mathbb{R}^n_v)) \cap L^\infty_{\mathrm{loc}}(\mathbb{R}_+; L^2(\mu)) \cap L^2_{\mathrm{loc}}(\mathbb{R}_+; H^1_v(\mu))$, *such that* $h(0,\cdot) = h_0$.

The proof of existence is a straightforward consequence of a standard approximation procedure together with the Helffer–Nier existence results, and the a priori estimate

$$\int h^2(t,x,v)\,d\mu(x,v) + \int_0^t \int h^2(s,x,v)\,d\mu(x,v)\,ds$$
$$\leq \int h^2(0,x,v)\,d\mu(x,v).$$

There is more to say about the uniqueness statement, of which the proof is deferred to Appendix A.20. The main subtlety lies in the absence of any growth condition on ∇V; this is overcome by a localization argument inspired from [**32**, Proposition 5.5]. Apart from that, Theorem 6 is just an exercise in linear partial differential equations.

Many people (including me) would rather think of (2.6) in the form

(2.7) $$\partial_t f + v \cdot \nabla_x f - \nabla V(x) \cdot \nabla_v f = \Delta_v f + \nabla_v \cdot (vf),$$

in which case f at time t can be interpreted (if it is nonnegative) as a density of particles, or (if it is a probability density) as the law of a random variable in phase space. To switch from (2.6) to (2.7) it suffices to set $f := f_\infty h$. This however does not completely solve the problem because the natural assumptions for (2.7) are much more general than for (2.6). For instance, it is natural to assume that the initial datum f_0 for (2.7) is L^2 with *polynomial* weight; or just L^1, or even a finite measure. Theorem 7 below yields a uniqueness result in such a setting, however with more stringent assumptions on the initial datum. In the next statement, $M(\mathbb{R}^n \times \mathbb{R}^n)$ stands for the space of finite measures on $\mathbb{R}^n \times \mathbb{R}^n$, equipped with the topology of weak convergence (against bounded continuous functions).

THEOREM 7. *Let* $V \in C^1(\mathbb{R}^n)$, $\inf V > -\infty$, *and let* $E(x,v) := V(x) + \frac{|v|^2}{2}$. *Then, for any* $f_0 \in L^2((1+E)\,dx\,dv)$, *equation* (2.7) *admits a unique distributional solution* $f = f(t,x,v) \in C(\mathbb{R}_+; \mathcal{D}'(\mathbb{R}^n_x \times \mathbb{R}^n_v)) \cap L^\infty_{\mathrm{loc}}(\mathbb{R}_+; L^2((1+E)\,dx\,dv)) \cap L^2_{\mathrm{loc}}(\mathbb{R}_+; H^1_v(\mathbb{R}^n_x \times \mathbb{R}^n_v))$, *such that* $f(0,\cdot) = f_0$.

If moreover $\nabla^2 V$ *is uniformly bounded, then for any finite measure* f_0 *the equation* (2.7) *admits a unique solution* $f = f(t,x,v) \in C(\mathbb{R}_+; M(\mathbb{R}^n_x \times \mathbb{R}^n_v))$.

The proof of this theorem will be deferred to Appendix A.20.

3. Coercivity and hypocoercivity

3.1. Coercivity.

DEFINITION 8. Let L be an unbounded operator on a Hilbert space $\mathcal{H}$, with kernel $\mathcal{K}$, and let $\widetilde{\mathcal{H}}$ be another Hilbert space continuously and densely embedded in $\mathcal{K}^\perp$, endowed with a scalar product $\langle \cdot, \cdot \rangle_{\widetilde{\mathcal{H}}}$ and a Hilbertian norm $\|\cdot\|_{\widetilde{\mathcal{H}}}$. The operator L is said to be λ-coercive on $\widetilde{\mathcal{H}}$ if

$$\forall h \in \mathcal{K}^\perp \cap D(L), \quad \Re \langle Lh, h \rangle_{\widetilde{\mathcal{H}}} \geq \lambda \|h\|_{\widetilde{\mathcal{H}}}^2,$$

where $\Re$ stands for real part. The operator L is said to be coercive on $\widetilde{\mathcal{H}}$ if it is λ-coercive on $\widetilde{\mathcal{H}}$ for some $\lambda > 0$.

The most standard situation is when $\widetilde{\mathcal{H}} = \mathcal{K}^\perp \simeq \mathcal{H}/\mathcal{K}$. Then it is equivalent to say that L is coercive on $\mathcal{K}^\perp$ (which will be abbreviated into just: L is coercive), or that the symmetric part of L admits a *spectral gap*.

Coercivity properties can classically be read at the level of the semigroup (assuming it is well-defined), as shown by the next statement:

PROPOSITION 9. *With the same notation as in Definition 8, L is λ-coercive on $\widetilde{\mathcal{H}}$ if and only if $\|e^{-tL}h_0\|_{\widetilde{\mathcal{H}}} \leq e^{-\lambda t} \|h_0\|_{\widetilde{\mathcal{H}}}$ for all $h_0 \in \widetilde{\mathcal{H}}$ and $t \geq 0$.*

PROOF. Assume by density that $h_0 \in \widetilde{\mathcal{H}} \cap D(L)$. On one hand the coercivity implies

$$\left.\frac{d}{dt}\right|_{t=0^+} \|e^{-tL}h_0\|_{\widetilde{\mathcal{H}}}^2 = -2\Re \langle Le^{-tL}h_0, e^{-tL}h_0 \rangle \leq -2\lambda \|e^{-tL}h_0\|^2,$$

so by Gronwall's lemma

$$\|e^{-tL}h_0\|_{\widetilde{\mathcal{H}}}^2 \leq e^{-2\lambda t}\|h_0\|_{\widetilde{\mathcal{H}}}^2.$$

Conversely, if exponential decay holds, then for any $h_0 \in \widetilde{\mathcal{H}} \cap D(L)$,

$$\Re \langle Lh_0, h_0 \rangle = \lim_{t \to 0} \frac{\|h_0\|_{\widetilde{\mathcal{H}}}^2 - \|e^{-tL}h_0\|_{\widetilde{\mathcal{H}}}^2}{2t}$$

$$\geq \liminf_{t \to 0} \frac{(1 - e^{-2\lambda t})\|h_0\|_{\widetilde{\mathcal{H}}}^2}{2t} = \lambda \|h_0\|^2,$$

whence the coercivity. □

When an operator L is in the form (2.1), the coercivity of L follows from the coercivity of A^*A, at least if B has a sufficiently large kernel:

PROPOSITION 10. *With the notation of Subsection 1.1, if A^*A is λ-coercive on $(\operatorname{Ker} A)^\perp$ and $\operatorname{Ker} A \subset \operatorname{Ker} B$, then L is λ-coercive on $\mathcal{K}^\perp$.*

PROOF. We know that $\mathcal{K} = \operatorname{Ker} A \cap \operatorname{Ker} B = \operatorname{Ker} A$, so for any $h \in \mathcal{K}^\perp$, $\langle Lh, h \rangle = \|Ah\|^2 \geq \lambda \|h\|^2$. □

EXAMPLE 11. Apart from trivial examples where $B = 0$, one can consider the following operator from [4]:

$$L = -(\Delta_x - x \cdot \nabla_x) - (\Delta_v - v \cdot \nabla_v) + (v \cdot \nabla_x - x \cdot \nabla_v)$$

on $L^2(e^{-(|v|^2 + |x|^2)/2}\, dx\, dv)$.

The main problem in the sequel is to study cases in which A^*A is coercive, but L is *not*, and yet there is exponential convergence to equilibrium (i.e. to an element of $\mathcal{K}$) for the semigroup (e^{-tL}). In view of Proposition 10, this can only happen if Ker L *is smaller than* Ker A. Here is the most typical example: on $\mathcal{H} = L^2(e^{-(|v|^2+|x|^2)/2}\, dv\, dx)$ again, consider

$$L = -(\Delta_v - v \cdot \nabla_v) + (v \cdot \nabla_x - x \cdot \nabla_v).$$

Then Ker A is made of functions which depend only on x, but Ker L only contains constants.

3.2. Hypocoercivity. To fix ideas, here is a (possibly misleading, but at least precise) definition of "hypocoercivity" in a Hilbertian context.

DEFINITION 12. Let $\mathcal{H}$ be a Hilbert space, L an unbounded operator on $\mathcal{H}$ generating a continuous semigroup $(e^{-tL})_{t\geq 0}$, and $\widetilde{\mathcal{H}}$ another Hilbert space, continuously and densely embedded in $\mathcal{K}^\perp$, endowed with a Hilbertian norm $\|\cdot\|_{\widetilde{\mathcal{H}}}$. The operator L is said to be λ-hypocoercive on $\widetilde{\mathcal{H}}$ if there exists a finite constant C such that

(3.1) $\qquad \forall h_0 \in \widetilde{\mathcal{H}}, \quad \forall t \geq 0 \quad \|e^{-tL}h_0\|_{\widetilde{\mathcal{H}}} \leq C e^{-\lambda t}\|h_0\|_{\widetilde{\mathcal{H}}}.$

It is said to be hypocoercive on $\widetilde{\mathcal{H}}$ if it is λ-hypocoercive on $\widetilde{\mathcal{H}}$ for some $\lambda > 0$.

REMARK 13. With respect to the definition of coercivity in terms of semigroups, the only difference lies in the appearance of the constant C in the right-hand side of (3.1) (obviously $C \geq 1$, apart from trivial cases; $C = 1$ would mean coercivity). The difference between Definition 8 and Definition 12 seems to be thin in view of the following fact (pointed to me by Serre): Whenever one has a norm satisfying inequality (3.1) for some constant C, it is always possible to find an equivalent norm (in general, not Hilbertian) for which the same inequality holds true with $C = 1$. Indeed, just choose

$$N(h) := \sup_{t \geq 0}\Big(e^{\lambda t}\,\|e^{-tL}h\|\Big).$$

In spite of these remarks, hypocoercivity is strictly weaker than coercivity. In particular, hypocoercivity is invariant under change of equivalent Hilbert norm on $\widetilde{\mathcal{H}}$, while coercivity is not. This has an important practical consequence: If one finds an equivalent norm for which the operator L is coercive, it follows that it is hypocoercive. I shall systematically use this strategy in the sequel.

REMARK 14. It often happens that a certain space $\widetilde{\mathcal{H}}$ is convenient for proving hypocoercivity, but this particular space is much smaller than $\mathcal{H}$ (stated otherwise, the Hilbert norm on $\widetilde{\mathcal{H}}$ cannot be bounded in terms of the Hilbert norm on $\mathcal{H}$): typically, $\widetilde{\mathcal{H}}$ may be a weighted Sobolev space, while $\mathcal{H}$ is a weighted L^2 space. In that situation there is in general no density argument which would allow one to go directly from hypocoercivity on $\widetilde{\mathcal{H}}$, to hypocoercivity on $\mathcal{H}$. However, such an extension is possible if L satisfies a (hypoelliptic) regularization estimate of the form

(3.2) $\qquad \forall t_0 > 0 \quad \exists C(t_0) < +\infty; \quad \forall t \geq t_0, \ \|e^{-tL}\|_{\mathcal{K}^\perp \to \widetilde{\mathcal{H}}} \leq C(t_0);$

or, more generally, if L generates a semigroup for which there is *exponential decay of singularities*:

(3.3)
$$\begin{cases} \forall t \geq 0, \quad e^{-tL} = S_t + R_t, \\ \forall t_0 > 0 \quad \exists C(t_0) < +\infty; \quad \forall t \geq t_0, \ \|S_t\|_{\mathcal{K}^\perp \to \widetilde{\mathcal{H}}} \leq C(t_0); \\ \exists \lambda > 0; \quad \forall t \geq 0, \ \|R_t\|_{\mathcal{K}^\perp \to \mathcal{K}^\perp} \leq C e^{-\lambda t}. \end{cases}$$

Such assumptions are often satisfied in realistic models. For instance, integral operators (generators of jump processes) usually satisfy (3.3) when the kernel is integrable (finite jump measure), and (3.2) when the kernel is not integrable. Diffusion operators of heat or Fokker–Planck type usually satisfy (3.2).

3.3. Commutators. If the operators A^*A and B commute, then so do their exponentials, and $e^{-tL} = e^{-tA^*A}e^{-tB}$. Then, since B is antisymmetric, e^{-tB} is norm-preserving, and it is equivalent to study the convergence for e^{-tL} or for e^{-tA^*A}. On the other hand, if these operators do *not* commute, one can hope for interesting phenomena.

PROPOSITION 15. *With the notation of Subsection* 1.1, *in particular* $L = A^*A + B$, *define recursively the iterated commutators*

$$C_0 := A, \quad C_k := [C_{k-1}, B],$$

and then $\mathcal{K}' := \cap_{k \geq 0} \operatorname{Ker} C_k$. *Then* $\mathcal{K} \subset \mathcal{K}'$, *and* $\mathcal{K}'$ *is invariant for* e^{-tL}.

PROOF. Assume that $\mathcal{K} \subset \operatorname{Ker} C_0 \cap \ldots \cap \operatorname{Ker} C_k$. Then, for all $h \in \mathcal{K} \cap \mathcal{S}$,

$$C_{k+1} h = C_k B h - B C_k h = C_k B h = -C_k A^*A h = 0.$$

Thus $\mathcal{K} \subset \operatorname{Ker} C_{k+1}$. By induction, $\mathcal{K}$ is included in the intersection $\mathcal{K}'$ of all $\operatorname{Ker} C_j$.

Next, if $h \in \mathcal{K}' \cap \mathcal{S}$, then $Lh = Bh$, so $C_k L h = C_k B h = C_{k+1} h + B C_k h = 0$; since k is arbitrary, in fact $Lh \in \mathcal{K}'$, so L leaves $\mathcal{K}'$ invariant, and therefore so does e^{-tL}. □

In most cases of interest, not only does $\mathcal{K}'$ coincide with $\mathcal{K}$, but in addition $\mathcal{K}'$ can be constructed as the intersection of just *finitely many* kernels of iterated commutators. Thanks to the trivial identity

$$\bigcap_{j=0}^{k} \operatorname{Ker} C_j = \operatorname{Ker} \left(\sum_{j=0}^{k} C_j^* C_j \right),$$

the condition that $\mathcal{K}'$ is the intersection of finitely many iterated commutators may be reformulated as follows:

(3.4) There exists $N_c \in \mathbb{N}$ such that $\operatorname{Ker} \left(\sum_{j=0}^{N_c} C_j^* C_j \right) = \operatorname{Ker} L.$

EXAMPLE 16. For the kinetic Fokker–Planck operator (2.6), $N_c = 1$ will do.

If the goal is to derive estimates on the rate of convergence, it is natural to reinforce the above condition into a more quantitative one:

(3.5) $$\sum_{k=0}^{N_c} C_k^* C_k \text{ is coercive on } \mathcal{K}^\perp.$$

Condition (3.5) is more or less an analogue of Hörmander's "rank r" bracket condition (as explained later, $r = 2N_c+1$ is the natural convention), but in the context of convergence to equilibrium and spectral gap, rather than regularization and elliptic estimates. There is however an important difference: Here we are taking brackets always with B, while in Hörmander's condition, brackets of the form, say, $[A_i, A_j]$ would be allowed. This modification is intentional: in all the cases of interest known to me, there is no need to consider such brackets for hypocoercivity problems. A basic example which will be discussed in Appendix A.19 is the following: The differential operator

$$L := -(x^2 \partial_y{}^* \partial_y + \partial_x{}^* \partial_x),$$

although *not* elliptic, is *coercive* (not just hypocoercive) in $L^2(\gamma)/\mathbb{R}$, where γ is the gaussian measure on $\mathbb{R}^2$. For this operator, brackets of the form $[\partial_x, x\partial_y]$ play a crucial role in the regularity study, but they are not needed to establish lower bounds on the spectral gap.

REMARK 17. It was pointed out to me by Serre that, when $\mathcal{H}$ is finite-dimensional, condition (3.5) is equivalent to the statement that Ker A *does not contain any nontrivial subspace invariant by B*. In the study of convergence to equilibrium for hyperbolic systems of conservation laws, this condition is known as **Kawashima's nondegeneracy condition** [31, 37, 50]. It is not so surprising to note that the very same condition appears in Hörmander's seminal 1967 paper on hypoellipticity [35, p. 148] as a necessary and sufficient condition for a diffusion equation to be hypoelliptic, when the constant matrices A and B respectively stand for the diffusion matrix and the linear drift function.[1]

Taking iterated commutators may rapidly lead to cumbersome expressions, because of "lower-order terms". In the present context, this might be more annoying than in a regularity context, and so it will be convenient to allow for perturbations in the definition of C_k, say

$$[C_k, B] = C_{k+1} + R_{k+1},$$

where R_{k+1} is a "remainder term", chosen according to the context, that is controlled by $C_0, \ldots, C_k$. An easy and sometimes useful generalization is to set

$$[C_k, B] = Z_{k+1} C_{k+1} + R_{k+1},$$

where the Z_k's are auxiliary operators, typically multipliers, satisfying certain identities.

Once the family $(C_0, \ldots, C_{N_c})$ is secured, one can introduce the corresponding abstract Sobolev $\mathcal{H}^1$ norm as in (1.1). This norm will be used on $\mathcal{H}$, or (more often)

---

[1] At first sight, it seems that both problems are completely different: Kawashima's condition is applied to systems of unknowns, while Hörmander's example deals with scalar equations. The analogy becomes less surprising when one notices that for such a diffusion equation the fundamental solution, viewed as a function of time, takes its values in the finite-dimensional space of Gaussian distributions, so that the equation really defines a system.

on $\mathcal{K}^\perp$. On the latter space we may also consider "homogeneous Sobolev norms" such as

(3.6) $$\|h\|_{\dot{\mathcal{H}}^1}^2 := \sum_{j=0}^{N_c} \|C_j h\|^2.$$

Note that, with the above assumptions, the orthogonal space to the kernel $\mathcal{K}$ in $\mathcal{H}^1$ *does not depend* on whether we consider the scalar product of $\mathcal{H}$ or that of $\mathcal{H}^1$. (See the proof of Theorem 24 below.) So a natural choice for $\widetilde{\mathcal{H}}$ will be

$$\widetilde{\mathcal{H}} = \mathcal{H}^1/\mathcal{K},$$

which is $\mathcal{K}^\perp$ equipped with the $\mathcal{H}^1$ norm.

4. Basic theorem

In this section linear operators satisfying a "rank-3" condition ($N_c = 1$ in (3.5)) are considered. Although this is a rather simple situation, it is already of interest, and its understanding will be the key to more complicated extensions; so I shall spend some time on this case. Here it will be assumed for pedagogical reasons that the operators A and C commute; this assumption will be relaxed in the next section.

THEOREM 18. *With the notation of Subsection 1.1, consider a linear operator $L = A^*A + B$ (B antisymmetric), and define $C := [A, B]$. Assume the existence of constants α, β such that*

(i) A and A^* commute with C; A commutes with A (i.e. each A_i commutes with each A_j);

(ii) $[A, A^*]$ is α-bounded relatively to I and A;

(iii) $[B, C]$ is β-bounded relatively to A, A^2, C and AC;

Then there is a scalar product $((\cdot, \cdot))$ on $\mathcal{H}^1/\mathcal{K}$, defining a norm equivalent to the $\mathcal{H}^1$ norm, such that

(4.1) $$\forall h \in \mathcal{H}^1/\mathcal{K}, \qquad ((h, Lh)) \geq K \left(\|Ah\|^2 + \|Ch\|^2 \right)$$

for some constant $K > 0$, only depending on α and β.

If, in addition,

$$A^*A + C^*C \text{ is } \kappa\text{-coercive}$$

for some $\kappa > 0$, then there is a constant $\lambda > 0$, only depending on α, β and κ, such that

$$\forall h \in \mathcal{H}^1/\mathcal{K}, \qquad ((h, Lh)) \geq \lambda((h, h)).$$

In particular, L is hypocoercive in $\mathcal{H}^1/\mathcal{K}$:

$$\|e^{-tL}\|_{\mathcal{H}^1/\mathcal{K} \to \mathcal{H}^1/\mathcal{K}} \leq c\, e^{-\lambda t} \qquad (c < +\infty),$$

where both λ and c can be estimated explicitly in terms of upper bounds on α and β, and a lower bound on κ.

Before stating the proof of Theorem 18, I shall provide some remarks and further explanations.

REMARK 19. Up to changing α and β, it is equivalent to impose (ii) and (iii) above or to impose the seemingly more general conditions:

(ii') $[A, A^*]$ is α-bounded relatively to I, A and A^*;

(iii') $[B, C]$ is β-bounded relatively to A, A^2, A^*A, C and AC.

Indeed,
$$\langle A^*h, A^*h \rangle = \langle AA^*h, h \rangle = \langle A^*Ah, h \rangle + \sum_i \langle [A_i, A_i^*]h, h \rangle,$$
so
$$\|A^*h\|^2 \leq \|Ah\|^2 + \|[A, A^*]h\| \, \|h\|.$$
Then assumption (ii') implies
$$\|A^*h\|^2 \leq \|Ah\|^2 + \alpha \Big(\|h\|^2 + \|Ah\| \, \|h\| + \|A^*h\| \, \|h\| \Big)$$
$$\leq \|Ah\|^2 + \alpha \Big(\|h\|^2 + \|Ah\| \, \|h\| \Big) + \frac{1}{2}\|A^*h\|^2 + \frac{\alpha^2}{2}\|h\|^2,$$
and it follows that A^* is bounded relatively to I and A, so that (ii) holds true. This also implies that A^*A is bounded relatively to A^2 and A, so (iii') implies (iii).

REMARK 20. Assumption (ii) in Theorem 18 can be relaxed into

(ii'') $[A, A^*]A$ is relatively bounded with respect to A and A^2; where by convention
$$\|[A, A^*]Ah\|^2 = \sum_i \Big\| \sum_j [A_i, A_j^*]A_j h \Big\|^2.$$

REMARK 21. Here is a crude heuristic rule helping to understand assumptions (i) to (iii) above. As is classical in Hörmander's theory, define the weights $w(O)$ of the operators involved, by
$$w(A) = w(A^*) = 1, \qquad w(B) = 2, \qquad w([O_1, O_2]) = w(O_1) + w(O_2).$$
Then rules (i) to (iii) guarantee that certain key commutators can be estimated in terms of operators whose order is strictly less: for instance, the weight of $[B, C]$ is $2+3=5$, and assumption (iii) states that it should be controlled by some operators, for which the maximal weight is 4. (This rule does not however explain why I is allowed in the right-hand side of (ii), but not in (iii); so it might be better to think in terms of Assumption (ii'') from Remark 20 rather than in terms of Assumption (ii).)

REMARK 22. In particular cases of interest, it may be a good idea to rewrite the proof of Theorem 18, taking into account specific features of the problem considered, so as to obtain better constants λ and C.

4.1. Heuristics and strategy. The proof of Theorem 18 is quite elementary; in some sense, the most sophisticated analytical tool on which it rests is the Cauchy–Schwarz inequality. The argument consists in devising an appropriate Hilbertian norm on $\mathcal{H}^1/\mathcal{K}$, which will be equivalent to the usual norm, but will turn L into a coercive operator. One can see an analogy with a classical, elementary proof of a standard theorem in linear algebra [**5**, pp. 147-148]: If the real parts of the eigenvalues of a matrix M are all positive, then $e^{-tM} \to 0$ (exponentially fast) as $t \to \infty$.

Define
$$(4.2) \qquad ((h, h)) = \|h\|^2 + a\,\|Ah\|^2 + 2b\,\Re\,\langle Ah, Ch \rangle + c\,\|Ch\|^2,$$
where the positive constants a, b, c will be chosen later on, in such a way that $1 \gg a \gg b \gg c$. (The constant c here is not the same as the one in the conclusion of Theorem 18.)

By polarization, this formula defines a bilinear symmetric form on $\mathcal{H}^1$. By using Young's inequality, in the form

$$\left|2b\left\langle Ah, Ch\right\rangle\right| \leq 2b\, \|Ah\|\, \|Ch\| \leq b\sqrt{\frac{a}{c}}\|Ah\|^2 + b\sqrt{\frac{c}{a}}\|Ch\|^2,$$

one sees that the scalar products $((\cdot,\cdot))$ and $\langle\cdot,\cdot\rangle_{\mathcal{H}^1}$ define equivalent norms as soon as $b < \sqrt{ac}$, and more precisely

$$(4.3) \quad \min(1,a,c)\left(1 - \frac{b}{\sqrt{ac}}\right)\|h\|_{\mathcal{H}^1}^2 \leq ((h,h))$$

$$\leq \max(1,a,c)\left(1 + \frac{b}{\sqrt{ac}}\right)\|h\|_{\mathcal{H}^1}^2.$$

In particular, the scalar products $((\cdot,\cdot))$ and $\langle\cdot,\cdot\rangle_{\mathcal{H}^1}$ define equivalent norms.

In spite of their equivalence, the scalar products $((\cdot,\cdot))$ and $\langle\cdot,\cdot\rangle_{\mathcal{H}^1}$ are quite different: it is possible to arrange that L is coercive with respect to the former, although it is not with respect to the latter. Heuristically, one may say that the "pure" terms $\|h\|^2$, $\|Ah\|^2$ and $\|Ch\|^2$ will mainly feel the influence of the symmetric part in L, but that the "mixed" term $\langle Ah, Ch\rangle$ will mainly feel the influence of the antisymmetric part in L. The following simple calculations should help understanding this. Whenever Q is a linear operator commuting with A (be it I, A or C in this example), one has

$$\left.\frac{d}{dt}\right|_{t=0} \|Qe^{-tA^*A}h\|^2 = -2\|QAh\|^2,$$

but on the other hand

$$\left.\frac{d}{dt}\right|_{t=0} \langle Ae^{-tB}h, Ce^{-tB}h\rangle = -\langle ABh, Ch\rangle - \langle Ah, CBh\rangle.$$

Pretend that B and C commute, and this can be rewritten

$$-\langle ABh, Ch\rangle - \langle Ah, BCh\rangle = -\langle ABh, Ch\rangle - \langle B^*Ah, Ch\rangle$$
$$= -\langle ABh, Ch\rangle + \langle BAh, Ch\rangle$$
$$= -\langle [A,B]h, Ch\rangle = -\|Ch\|^2,$$

where the antisymmetry of B has been used to go from the first to the second line. This will yield the dissipation in the C direction, which the symmetric part of A was unable to provide!

4.2. Proof of Theorem 18. Introduce the norm (4.2). By Proposition 15, any $h \in \mathcal{K} = \operatorname{Ker} L$ satisfies $Ah = 0$, $Ch = 0$, in which case $((h,h')) = \langle h,h'\rangle_{\mathcal{H}^1} = \langle h, h'\rangle$ for all $h' \in \mathcal{H}$. In particular, the orthogonal space $\mathcal{K}^\perp$ is the same for these three scalar products. So it makes sense to choose $\widetilde{\mathcal{H}} = \mathcal{H}^1/\mathcal{K}$.

Let us compute

$$-\frac{1}{2}\frac{d}{dt}\left((e^{-tL}h, e^{-tL}h)\right) = \Re\left((e^{-tL}h, Le^{-tL}h)\right);$$

if we can bound below this time-derivative by a constant multiple of $((e^{-tL}h, e^{-tL}h))$, then the conclusion of Theorem 18 will follow by Gronwall's lemma. By semigroup property, it is sufficient to consider $t=0$, so the problem is to bound below

$\Re((h, Lh))$ by a multiple of $((h, h))$. Obviously,

(4.4) $$\Re\left((h, Lh)\right) = \Re\langle h, Lh\rangle + a\,(\mathrm{I}) + b\,(\mathrm{II}) + c\,(\mathrm{III}),$$

where

$$(\mathrm{I}) := \Re\langle Ah, ALh\rangle, \qquad (\mathrm{II}) := \Re\langle ALh, Ch\rangle + \Re\langle Ah, CLh\rangle,$$
$$(\mathrm{III}) := \Re\langle Ch, CLh\rangle.$$

By Proposition 2(i), $\Re\langle h, Lh\rangle = \|Ah\|^2$. For each of the terms (I), (II), (III), the contributions of A^*A and B will be estimated separately, and the resulting expressions will be denoted $(\mathrm{I})_A$, $(\mathrm{I})_B$, $(\mathrm{II})_A$, $(\mathrm{II})_B$, etc. For consistency with the sequel, I shall introduce the notation

(4.5) $$R_2 := [C, B].$$

Moreover, to alleviate notation, I shall temporarily assume that $\mathcal{H}$ is a real Hilbert space; otherwise, just put real parts everywhere.

First of all,

$$(\mathrm{I})_B = \langle Ah, ABh\rangle = \langle Ah, BAh\rangle + \langle Ah, [A, B]h\rangle$$
$$= 0 + \langle Ah, Ch\rangle \geq -\|Ah\|\|Ch\|,$$

where the antisymmetry of B was used. Then,

$$(\mathrm{I})_A = \langle Ah, AA^*Ah\rangle = \langle A^2 h, A^2 h\rangle + \langle Ah, [A, A^*]Ah\rangle,$$

to be understood as

$$\sum_{ij} \langle A_j A_i h, A_i A_j h\rangle + \langle A_i h, [A_i, A_j^*]A_j h\rangle.$$

This can be rewritten

$$\sum_{ij} \|A_i A_j h\|^2 + \langle [A_j, A_i]h, A_i A_j h\rangle + \langle A_i h, [A_i, A_j^*]A_j h\rangle$$
$$\equiv \|A^2 h\|^2 + \langle [A, A]h, A^2 h\rangle + \langle Ah, [A, A^*]Ah\rangle.$$

In the present case it is assumed that $[A, A] = 0$, so the second term vanishes. Then from the Cauchy–Schwarz inequality we have

$$(\mathrm{I})_A \geq \|A^2 h\|^2 - \|Ah\|\,\|[A, A^*]Ah\|.$$

4. BASIC THEOREM

Next,
$$\begin{aligned}(II)_B &= \langle ABh, Ch\rangle + \langle Ah, CBh\rangle \\ &= \langle ABh, Ch\rangle + \langle Ah, BCh\rangle + \langle Ah, [C,B]h\rangle \\ &= \langle ABh, Ch\rangle - \langle BAh, Ch\rangle + \langle Ah, R_2 h\rangle \\ &= \langle [A,B]h, Ch\rangle + \langle Ah, R_2 h\rangle \\ &\geq \|Ch\|^2 - \|Ah\|\,\|R_2 h\|; \\ (II)_A &= \langle Ah, CA^*Ah\rangle + \langle AA^*Ah, Ch\rangle \\ &= \langle Ah, A^*CAh\rangle + \langle A^*A^2 h, Ch\rangle + \langle [A,A^*]Ah, Ch\rangle \\ &= \langle A^2 h, CAh\rangle + \langle A^2 h, ACh\rangle + \langle [A,A^*]Ah, Ch\rangle \\ &= 2\langle A^2 h, CAh\rangle + \langle Ch, [A,A^*]Ah\rangle \\ &\geq -2\|A^2 h\|\,\|CAh\| - \|Ch\|\,\|[A,A^*]Ah\|.\end{aligned}$$

(Here the commutation of C with both A and A^* was used.)

Finally,
$$\begin{aligned}(III)_B &= \langle Ch, CBh\rangle = \langle Ch, BCh\rangle + \langle Ch, [C,B]h\rangle \\ &= 0 + \langle Ch, R_2 h\rangle \\ &\geq -\|Ch\|\,\|R_2 h\|; \\ (III)_A &= \langle Ch, CA^*Ah\rangle = \langle Ch, A^*CAh\rangle = \langle ACh, CAh\rangle = \|CAh\|^2\end{aligned}$$

(here again the commutation of C with A and A^* was used).

On the whole,

$$\begin{aligned}(4.6)\quad \Re\left((h, Lh)\right) \geq\; & \|Ah\|^2 \\ &+ a\Big(\|A^2 h\|^2 - \|Ah\|\,\|[A,A^*]Ah\| - \|Ah\|\,\|Ch\|\Big) \\ &+ b\Big(\|Ch\|^2 - \|Ah\|\,\|R_2 h\| - 2\|A^2 h\|\,\|CAh\| - \|Ch\|\,\|[A,A^*]Ah\|\Big) \\ &+ c\Big(\|CAh\|^2 - \|Ch\|\,\|R_2 h\|\Big).\end{aligned}$$

The assumptions of Theorem 18 imply
$$\|[A,A^*]y\| \leq \alpha\big(\|y\| + \|Ay\|\big),$$
$$\|R_2 h\| \leq \beta\big(\|Ah\| + \|A^2 h\| + \|Ch\| + \|CAh\|\big).$$

Plugging this into (4.6), follows an estimate which can be conveniently recast as
$$\Re\left((h, Lh)\right) \geq \langle X, mX\rangle_{\mathbb{R}^4},$$
where X is a vector in $\mathbb{R}^4$ and $m = [m_{ij}]_{1\leq i,j\leq 4}$ is a 4×4 matrix, say upper-diagonal:

$$X := \Big(\|Ah\|, \|A^2 h\|, \|Ch\|, \|CAh\|\Big),$$

$$m := \begin{bmatrix} 1-(a\alpha + b\beta) & -(a\alpha + b\beta) & -(a + b\alpha + b\beta + c\beta) & -b\beta \\ 0 & a & -(b\alpha + c\beta) & -2b \\ 0 & 0 & b - c\beta & -c\beta \\ 0 & 0 & 0 & c \end{bmatrix}.$$

If the symmetric part of m is definite positive, this will imply inequality (4.1). Then the rest of Theorem 18 follows easily, since the κ-coercivity of $A^*A + C^*C$ implies

$$\|Ah\|^2 + \|Ch\|^2 \geq \frac{1}{2}(\|Ah\|^2 + \|Ch\|^2) + \frac{\kappa}{2}\|h\|^2$$
$$\geq \frac{\min(1,\kappa)}{2}\|h\|^2_{\mathcal{H}^1}.$$

So it all boils down now to choosing the parameters a, b and c in such a way that the symmetric part of m is positive definite, and for this it is sufficient to ensure that

$$\begin{cases} \forall i, \quad m_{ii} > 0; \\ \forall (i,j), \quad i \neq j \Longrightarrow m_{ij} \ll \sqrt{m_{ii}m_{jj}}. \end{cases}$$

In the sequel, the statement "the symmetric part of m_1 is greater than the symmetric part of m_2" will be abbreviated into just "m_1 is greater than m_2".

Let $M := \max(1, \alpha, \beta)$. Assume, to fix ideas, that

(4.7) $$1 \geq a \geq b \geq 2c.$$

Then m can be bounded below (componentwise) by

$$\begin{bmatrix} 1-2Ma & -2Ma & -4Ma & -Mb \\ 0 & a & -2Mb & -2Mb \\ 0 & 0 & b-Mc & -Mc \\ 0 & 0 & 0 & c \end{bmatrix}.$$

If now it is further assumed that

(4.8) $$a \leq \frac{1}{4M}, \quad c \leq \frac{b}{2M},$$

then the latter matrix can in turn be bounded below by

$$\begin{bmatrix} 1/2 & -2Ma & -4Ma & -Mb \\ 0 & a & -2Mb & -2Mb \\ 0 & 0 & b/2 & -Mc \\ 0 & 0 & 0 & c \end{bmatrix} \equiv [\widetilde{m}_{ij}].$$

By imposing

(4.9) $$|\widetilde{m}_{ij}| \leq \sqrt{\widetilde{m}_{ii}\widetilde{m}_{jj}}/2 \leq (\widetilde{m}_{ii} + \widetilde{m}_{jj})/4,$$

it will follow

$$\sum_{ij} \widetilde{m}_{ij} X_i X_j \geq \sum_i \widetilde{m}_{ii} X_i^2 - \frac{3}{4}\sum_i \widetilde{m}_{ii} X_i^2 = \frac{1}{4}\sum \widetilde{m}_{ii} X_i^2.$$

(The 3 in 3/4 is because each diagonal term should participate in the control of three off-diagonal terms.) To ensure (4.9), it suffices that

$$2Ma \leq \sqrt{\frac{a}{8}}, \quad 4Ma \leq \sqrt{\frac{b}{16}}, \quad Mb \leq \sqrt{\frac{c}{8}}, \quad 2Mb \leq \sqrt{\frac{ab}{8}},$$

$$2Mb \leq \sqrt{\frac{ac}{4}}, \quad Mc \leq \sqrt{\frac{bc}{8}}.$$

All these conditions, including (4.8), are fulfilled if

(4.10) $$a, \frac{b}{a}, \frac{c}{b} \leq \frac{1}{32\,M^2} \qquad \frac{a^2}{b}, \frac{b^2}{ac} \leq \frac{1}{256\,M^2}.$$

Lemma A.22 in Appendix A.23 shows that it is always possible to choose a, b, c in such a way. This concludes the proof of Theorem 18. □

REMARK 23. There are other possible ways to conduct these calculations. In an early version of this work, the last term $(III)_A$ was rewritten in three different forms to create helpful terms in $\|AC^*h\|^2$ and $\|A^*Ch\|^2$, at the cost of requiring additional assumptions on $[C, C^*]$.

5. Generalization

Now I shall present a variant of Theorem 18 which covers more general situations.

THEOREM 24. *Let $\mathcal{H}$ be a Hilbert space, let $A : \mathcal{H} \to \mathcal{H}^n$ and $B : \mathcal{H} \to \mathcal{H}$ be unbounded operators, $B^* = -B$, let $L := A^*A + B$ and $\mathcal{K} := \mathrm{Ker}\, L$. Assume the existence of $N_c \in \mathbb{N}$ and (possibly unbounded) operators $C_0, C_1, \ldots, C_{N_c+1}$, $R_1, \ldots, R_{N_c+1}$ and $Z_1, \ldots, Z_{N_c+1}$ such that*

$$C_0 = A, \qquad [C_j, B] = Z_{j+1}C_{j+1} + R_{j+1} \quad (0 \leq j \leq N_c), \qquad C_{N_c+1} = 0,$$

and, for all $k \in \{0, \ldots, N_c\}$,

  (i) $[A, C_k]$ is bounded relatively to $\{C_j\}_{0 \leq j \leq k}$ and $\{C_j A\}_{0 \leq j \leq k-1}$;
  (ii) $[C_k, A^*]$ is bounded relatively to I and $\{C_j\}_{0 \leq j \leq k}$;
  (iii) R_k is bounded relatively to $\{C_j\}_{0 \leq j \leq k-1}$ and $\{C_j A\}_{0 \leq j \leq k-1}$;
  (iv) There are positive constants λ_j, Λ_j such that $\lambda_j I \leq Z_j \leq \Lambda_j I$.

Then there is a scalar product $((\cdot, \cdot))$ on $\mathcal{H}^1$, defining a norm equivalent to the $\mathcal{H}^1$ norm,

$$\|h\|_{\mathcal{H}^1} := \sqrt{\|h\|^2 + \sum_{k=0}^{N_c} \|C_k h\|^2},$$

such that

(5.1) $$\forall h \in \mathcal{H}^1/\mathcal{K}, \qquad \Re\,((h, Lh)) \geq K \sum_{j=0}^{N_c} \|C_j h\|^2$$

for some constant $K > 0$, only depending on the bounds appearing implicitly in assumptions (i)–(iv).

If, in addition,

$$\sum_{j=0}^{N_c} C_j^* C_j \ \text{is}\ \kappa\text{-coercive}$$

for some $\kappa > 0$, then there is a constant $\lambda > 0$, only depending on K and κ, such that

$$\forall h \in \mathcal{H}^1/\mathcal{K}, \qquad \Re\,((h, Lh)) \geq \lambda\,((h, h)).$$

In particular, L is hypocoercive in $\mathcal{H}^1/\mathcal{K}$: There are constants $C \geq 0$ and $\lambda > 0$, explicitly computable in terms of the bounds appearing implicitly in assumptions (i)–(iii), and κ, such that

$$\|e^{-tL}\|_{\mathcal{H}^1/\mathcal{K} \to \mathcal{H}^1/\mathcal{K}} \leq C e^{-\lambda t}.$$

This result generalizes Theorem 18 in several respects: successive commutators are allowed, remainders R_{j+1} and multiplicators Z_{j+1} are allowed in the identity defining C_{j+1} in terms of C_j, and the operators $C_0, \ldots, C_{N_c}$ are not assumed to commute.

REMARK 25. The same rule as in Remark 21 applies to Assumptions (i)–(iii).

REMARK 26. Theorem A.12 in Appendix A.21 will show that the same structure assumptions (i)–(iv) imply an immediate regularization effect $\mathcal{H} \to \mathcal{H}^1$. This extends the range of application of the method, allowing data which do not necessarily lie in $\mathcal{H}^1$ but only in $\mathcal{H}$.

PROOF OF THEOREM 24. It is an amplification of the proof of Theorem 18. Let

$$(5.2) \qquad ((h,h)) := \|h\|^2 + \sum_{k=0}^{N_c} \Big(a_k \|C_k h\|^2 + 2\Re\, b_k \langle C_k h, C_{k+1} h \rangle \Big),$$

where $\{a_k\}_{0 \leq k \leq N_c+1}$ and $\{b_k\}_{0 \leq k \leq N_c}$ are families of positive coefficients, satisfying

$$(5.3) \qquad 0 \leq k \leq N_c \Longrightarrow \begin{cases} a_0 \leq \delta, \qquad b_k \leq \delta\, a_k, \qquad a_{k+1} \leq \delta\, b_k, \\ a_k^2 \leq \delta\, b_{k-1} b_k \quad (1 \leq k \leq N_c), \\ b_k^2 \leq \delta\, a_k a_{k+1} \quad (0 \leq k \leq N_c). \end{cases}$$

The small number $\delta > 0$ will be chosen later on, and the existence of the coefficients a_k, b_k is guaranteed by Lemma A.22 again.

Since $C_{N_c+1} = 0$, the last term in (5.2), with coefficient b_{N_c}, does not play any role. For $k \leq N_c - 1$, the inequality $b_k \leq \delta \sqrt{a_k a_{k+1}}$ implies

$$b_k \big| \langle C_k h, C_{k+1} h \rangle \big| \leq \frac{\delta a_k}{2} \|C_k h\|^2 + \frac{\delta a_{k+1}}{2} \|C_{k+1} h\|^2.$$

Hence, for δ small enough,

$$((h,h)) \geq \|h\|^2 + \frac{1}{2} \sum_{k=0}^{N_c} a_k \|C_k h\|^2,$$

so the norm defined by (5.2) is indeed equivalent to the $\mathcal{H}^1$ norm.

The next observation is that the space $\mathcal{K}^\perp$ is the same, whether the orthogonality is defined with respect to the scalar product in $\mathcal{H}$, the one in $\mathcal{H}^1$ or the one defined by (5.2). To show this it is sufficient to prove

$$(5.4) \qquad h \in \operatorname{Ker} L \Longrightarrow \quad \forall k \in \{0, \ldots, N_c+1\}, \quad C_k h = 0.$$

This will be achieved by finite induction on k. Let $h \in \operatorname{Ker} L$; by Lemma 2, $Ah = 0$, $Bh = 0$; so (5.4) is true for $k = 0$. Assume now that $C_j h = 0$ for $j \leq k$; it is obvious that also $C_j A h = 0$ for $j \leq k$; then our assumption on R_{k+1} implies that $R_{k+1} h = 0$. So $C_{k+1} h = C_k B h - B C_k h - R_{k+1} h = 0$. This concludes the proof of (5.4).

To prove (5.1) it is obviously sufficient to establish

$$(5.5) \qquad \Re\,((h, Lh)) \geq \frac{1}{2} \|Ah\|^2 + \sum_{k=0}^{N_c} \Big(\frac{a_k}{2} \|C_k A h\|^2 + \frac{b_k}{2} \|C_{k+1} h\|^2 \Big).$$

5. GENERALIZATION

As in the proof of Theorem 18 one can compute, with obvious notation,

$$\text{(5.6)} \qquad \Re\left((h, Lh)\right) = \|Ah\|^2 + \sum_{k=0}^{N_c} \Big\{ a_k [(\text{I})_A^k + (\text{I})_B^k] + 2 b_k [(\text{II})_A^k + (\text{II})_B^k] \Big\}.$$

To alleviate the notation, assume for a moment that we are working in a real Hilbert space, so there is no need to take real parts (otherwise, just put real parts everywhere). Explicit computations yield, for any $k \leq N_c$,

$$\begin{aligned}
(\text{I})_B^k &= \langle C_k h, C_k B h \rangle = \langle C_k h, [C_k, B] h \rangle + \langle C_k h, B C_k h \rangle \\
&= \langle C_k h, [C_k, B] h \rangle + 0 \\
&= \langle C_k h, Z_{k+1} C_{k+1} h \rangle + \langle C_k h, R_{k+1} h \rangle + 0 \\
&\geq -\Lambda_{k+1} \|C_k h\| \|C_{k+1} h\| - \|C_k h\| \|R_{k+1} h\| \\
(\text{I})_A^k &= \langle C_k h, C_k A^* A h \rangle \\
&= \langle C_k h, A^* C_k A h \rangle + \langle C_k h, [C_k, A^*] A h \rangle \\
&= \langle A C_k h, C_k A h \rangle + \langle C_k h, [C_k, A^*] A h \rangle \\
&= \langle C_k A h, C_k A h \rangle + \langle [A, C_k] h, C_k A h \rangle + \langle C_k h, [C_k, A^*] A h \rangle \\
&\geq \|C_k A h\|^2 - \|C_k A h\| \|[A, C_k] h\| - \|C_k h\| \|[C_k, A^*] A h\|
\end{aligned}$$

and, for $k \leq N_c - 1$ (when $k = N_c$, $(\text{II})_B^k = 0$),

$$\begin{aligned}
(\text{II})_B^k &= \langle C_k B h, C_{k+1} h \rangle + \langle C_k h, C_{k+1} B h \rangle \\
&= \langle C_k B h, C_{k+1} h \rangle + \langle C_k h, B C_{k+1} h \rangle + \langle C_k h, [C_{k+1}, B] h \rangle \\
&= \langle C_k B h, C_{k+1} h \rangle - \langle C_k h, B C_{k+1} h \rangle + \langle C_k h, [C_{k+1}, B] h \rangle \\
&= \langle [C_k, B] h, C_{k+1} h \rangle + \langle C_k h, [C_{k+1}, B] h \rangle \\
&= \langle C_k B h, C_{k+1} h \rangle - \langle B C_k h, C_{k+1} h \rangle + \langle C_k h, Z_{k+2} C_{k+2} h \rangle \\
&\qquad\qquad\qquad + \langle C_k h, R_{k+2} h \rangle \\
&= \langle [C_k, B] h, C_{k+1} h \rangle + \langle C_k h, Z_{k+2} C_{k+2} h \rangle + \langle C_k h, R_{k+2} h \rangle \\
&= \langle Z_{k+1} C_{k+1} h, C_{k+1} h \rangle + \langle R_{k+1} h, C_{k+1} h \rangle + \langle C_k h, Z_{k+2} C_{k+2} h \rangle \\
&\qquad\qquad\qquad + \langle C_k h, R_{k+2} h \rangle \\
&\geq \lambda_{k+1} \|C_{k+1} h\|^2 - \|C_{k+1} h\| \|R_{k+1} h\| - \Lambda_{k+2} \|C_k h\| \|C_{k+2} h\| \\
&\qquad\qquad\qquad - \|C_k h\| \|R_{k+2} h\|
\end{aligned}$$

$$(\text{II})_A^k = \langle C_k h, C_{k+1} A^* A h \rangle + \langle C_k A^* A h, C_{k+1} h \rangle$$
$$= \langle C_k h, [C_{k+1}, A^*] A h \rangle + \langle C_k h, A^* C_{k+1} A h \rangle + \langle A^* C_k A h, C_{k+1} h \rangle$$
$$+ \langle [C_k, A^*] A h, C_{k+1} h \rangle$$
$$= \langle C_k h, [C_{k+1}, A^*] A h \rangle + \langle A C_k h, C_{k+1} A h \rangle + \langle C_k A h, A C_{k+1} h \rangle$$
$$+ \langle [C_k, A^*] A h, C_{k+1} h \rangle$$
$$= \langle C_k h, [C_{k+1}, A^*] A h \rangle + \langle C_k A h, C_{k+1} A h \rangle + \langle [A, C_k] h, C_{k+1} A h \rangle$$
$$+ \langle C_k A h, C_{k+1} A h \rangle + \langle C_k A h, [A, C_{k+1}] h \rangle + \langle [C_k, A^*] A h, C_{k+1} h \rangle$$
$$\geq -\|C_k h\| \, \|[C_{k+1}, A^*] A h\| - \|C_k A h\| \, \|C_{k+1} A h\|$$
$$- \|C_{k+1} A h\| \, \|[A, C_k] h\| - \|C_k A h\| \, \|C_{k+1} A h\|$$
$$- \|C_k A h\| \, \|[A, C_{k+1}] h\| - \|C_{k+1} h\| \, \|[C_k, A^*] A h\|.$$

The next step is to use the quantities $a_j \|C_j A h\|^2$ and $b_j \|C_{j+1} h\|^2$ to control all the remaining terms. For this I shall apply Young's inequality, in the form $XY \leq \varepsilon X^2 + C(\varepsilon) Y^2$ ($C(\varepsilon) = \varepsilon^{-1}/4$). In the computations below, the dependence of C on the constants Λ_j will not be recalled.

$$(5.7) \qquad a_k [(\text{I})_A^k + (\text{I})_B^k] + b_k [(\text{II})_A^k + (\text{II})_B^k] \geq a_k \|C_k A h\|^2 + b_k \lambda_{k+1} \|C_{k+1} h\|^2$$

$$(5.8) \qquad -\varepsilon b_{k-1} \|C_k h\|^2 - C \frac{a_k^2}{b_{k-1}} \|C_{k+1} h\|^2$$

$$(5.9) \qquad -\varepsilon b_{k-1} \|C_k h\|^2 - C \frac{a_k^2}{b_{k-1}} \|R_{k+1} h\|^2$$

$$(5.10) \qquad -\varepsilon a_k \|C_k A h\|^2 - C a_k \|[A, C_k] h\|^2$$

$$(5.11) \qquad -\varepsilon b_{k-1} \|C_k h\|^2 - C \frac{a_k^2}{b_{k-1}} \|[C_k, A^*] A h\|^2$$

$$(5.12) \qquad -\varepsilon b_k \|C_{k+1} h\|^2 - C b_k \|R_{k+1} h\|^2$$

$$(5.13) \qquad -\varepsilon b_{k-1} \|C_k h\|^2 - C \frac{b_k^2}{b_{k-1}} \|C_{k+2} h\|^2$$

$$(5.14) \qquad -\varepsilon b_{k-1} \|C_k h\|^2 - C \frac{b_k^2}{b_{k-1}} \|R_{k+2} h\|^2$$

$$(5.15) \qquad -\varepsilon b_{k-1} \|C_k h\|^2 - C \frac{b_k^2}{b_{k-1}} \|[C_{k+1}, A^*] A h\|^2$$

$$(5.16) \qquad -\varepsilon a_k \|C_k A h\|^2 - C \frac{b_k^2}{a_k} \|C_{k+1} A h\|^2$$

$$(5.17) \qquad -\varepsilon a_{k+1} \|C_{k+1} A h\|^2 - C \frac{b_k^2}{a_{k+1}} \|[A, C_k] h\|^2$$

$$(5.18) \qquad -\varepsilon a_k \|C_k A h\|^2 - C \frac{b_k^2}{a_k} \|C_{k+1} A h\|^2$$

$$(5.19) \qquad -\varepsilon a_k \|C_k A h\|^2 - C \frac{b_k^2}{a_k} \|[A, C_{k+1}] h\|^2$$

$$(5.20) \qquad -\varepsilon b_k \|C_{k+1} h\|^2 - C \frac{b_k^2}{a_k} \|[C_k, A^*] A h\|^2,$$

5. GENERALIZATION

with the understanding that (5.12) to (5.20) in the above are not present when $k = N_c$. The problem is to show that each of the terms appearing in lines (5.8) to (5.20) can be bounded below by

$$(5.21) \qquad -\varepsilon\Big(\|Ah\|^2 + \sum_j (a_j \|C_j Ah\|^2 + b_j \|C_{j+1} h\|^2)\Big),$$

as soon as δ is small enough. This is true, by construction, of all the terms appearing on the left in these lines; so let us see how to control all the terms on the right. In the sequel, the notation $u \ll v$ means $u \leq \eta v$, where $\eta > 0$ becomes arbitrarily small as $\delta \to 0$.

- For lines (5.8), (5.13), (5.16) and (5.18), it is sufficient to impose

$$\frac{a_k^2}{b_{k-1}} \ll b_k, \qquad \frac{b_k^2}{b_{k-1}} \ll b_{k+1}, \qquad \frac{b_k^2}{a_k} \ll a_{k+1}.$$

The first and the third of these inequalities are true by construction; as for the second one, it follows from

$$b_k^2 \ll a_k a_{k+1} \ll \sqrt{b_{k-1} b_k} \sqrt{b_k b_{k+1}} \implies b_k^2 \ll b_{k-1} b_{k+1}.$$

- For lines (5.9), (5.12) and (5.14), we know that $\|R_{k+1} h\|$ is controlled by a combination of $\|Ah\|$, $\|A^2 h\|$, $\|C_1 h\|$, $\|C_1 Ah\|$, ..., $\|C_k h\|$, $\|C_k Ah\|$; hence it is sufficient to bound the coefficients appearing in front of $R_{k+1} h$ (resp. $R_{k+2} h$) by a small multiple of a_k (resp. a_{k+1}). So these terms are fine as soon as

$$\frac{a_k^2}{b_{k-1}} \ll a_k, \qquad b_k \ll a_k, \qquad \frac{b_k^2}{b_{k-1}} \ll a_{k+1}.$$

The second of these inequalities is true by construction, while the first and third one follow from

$$a_k^2 \ll b_{k-1} b_k \ll b_{k-1} a_k, \qquad b_k^2 \ll b_{k-1} b_{k+1} \ll b_{k-1} a_{k+1}.$$

- For lines (5.10), (5.17) and (5.19), we know that $\|[A, C_k]h\|$ is controlled by $\|Ah\|$, $\|A^2 h\|$, $\|C_1 h\|$, $\|C_1 Ah\|$, ..., $\|C_k h\|$. By a reasoning similar to the one above, it is sufficient to ensure

$$a_k \ll b_{k-1}, \qquad \frac{b_k^2}{a_{k+1}} \ll b_{k-1}, \qquad \frac{b_k^2}{a_k} \ll b_k.$$

The first and third of these inequalities are true by construction, while the second one follows from

$$b_k^2 \ll a_k a_{k+1} \ll b_{k-1} a_{k+1}.$$

- For lines (5.11), (5.15) and (5.20), we know that $\|[C_k, A^*]y\|$ is controlled by $\|y\|$, $\|Ay\|$, $\|C_1 y\|$, ..., $\|C_k y\|$, so $\|[C_k, A^*]Ah\|$ is controlled by $\|Ah\|$, $\|A^2 h\|$, $\|C_1 Ah\|$, ..., $\|C_k Ah\|$. By a reasoning similar to the one above, it is sufficient to ensure

$$\frac{a_k^2}{b_{k-1}} \ll a_k, \qquad \frac{b_k^2}{b_{k-1}} \ll a_{k+1}, \qquad \frac{b_k^2}{a_k} \ll a_k.$$

The first and third of these inequalities are true by construction, while the second one follows from

$$b_k^2 \ll a_k a_{k+1} \ll b_{k-1} a_{k+1}.$$

Putting all together, for all η there is a δ such that each of the "error terms" which appeared in the estimates above can be bounded below by (5.21). Then, if N_e stands for the number of error terms,

$$\Re\left((h, Lh)\right) \geq \left(1 - N_e \varepsilon\right) \left(\|Ah\|^2 + \sum_{k=0}^{N_c} \left(\frac{a_k}{2} \|C_k Ah\|^2 + \frac{b_k}{2} \|C_{k+1} h\|^2 \right) \right),$$

which implies (5.5) for δ small enough. The proof of (5.1) is now complete, and the end of Theorem 24 follows easily, as in the proof of Theorem 18. □

I shall conclude this section with a simple generalization of Theorem 24:

THEOREM 27. *With the same notation as in Theorem 24, define $C_{j+1/2} = C_j A$. Then the conclusion of Theorem 24 still holds true if Assumptions (i) to (iii) are relaxed as follows: There exists a constant M such that*
 (i') $\|[A, C_k]h\| \leq M \sum \|C_\alpha h\|^\theta \|C_\beta h\|^{1-\theta}$, *where the sum is over all pairs of indices (α, β) such that $\theta \alpha + (1-\theta)\beta \leq k + 1/2$, $(\alpha, \beta) \neq (k+1/2, k+1/2)$;*
 (ii') $\|[C_k, A^*]Ah\| \leq M \sum \|C_\alpha h\|^\theta \|C_\beta h\|^{1-\theta}$, *where the sum is over all pairs of indices (α, β) such that $\theta \alpha + (1-\theta)\beta \leq k + 1$, $(\alpha, \beta) \neq (k+1, k+1)$;*
 (iii') $\|R_k h\| \leq M \sum \|C_\alpha h\|^\theta \|C_\beta h\|^{1-\theta}$, *where the sum is over all pairs of indices (α, β) such that $\theta \alpha + (1-\theta)\beta \leq k$, $(\alpha, \beta) \neq (k, k)$.*

In statements (i') to (iii'), θ may vary from one pair (α, β) to the other. The conditions on admissible pairs (α, β) can be understood more easily if one remembers Remark 21; then the weight $w(C_\alpha)$ is $2\alpha + 1$. If one formally attributes to $[A, C_k]$, $[A^*, C_k]A$ and R_k the weights $2k+2$, $2k+3$ and $2k+1$, and decides that the weight of a formal product $\|O_1 h\|^\theta \|O_2 h\|^{1-\theta}$ is $\theta w(O_1) + (1-\theta) w(O_2)$, then Conditions (i') to (iii') mean that each of the operators $[A, C_k]$, $[A^*, C_k]A$ and R_k can be bounded in terms of lower weights. For instance, an estimate like

$$\|R_5 h\| \leq M \sqrt{\|C_3 h\| \|C_5 h\|}$$

is admissible, since $[w(C_3) + w(C_5)]/2 = (7+11)/2 < 11 = w(R_5)$.

PROOF OF THEOREM 27. The strategy is the same as in Theorem 24; but now one should use Young's inequality in the form

$$a^\theta b^{1-\theta} \leq \varepsilon a + C b,$$

and note that, if (u_m) is given by Lemma A.22, then

$$\left[\ell \leq \frac{m+n}{2}, \quad (m,n) \neq (\ell, \ell) \right] \implies u_\ell \ll \sqrt{u_m u_n}.$$

Then all the estimates entering the proof of Theorem 24 can be adapted without difficulty. □

6. Hypocoercivity in entropic sense

In this section I shall consider the problem of convergence to equilibrium for solutions of diffusion equations in an $L \log L$ setting. This represents a significant extension of the results already discussed, because in many cases of interest, after a finite time the solution automatically belongs to $L \log L(\mu)$, where μ is the stationary solution, but not to $L^2(\mu)$.[2]

For that purpose, I shall use the same information-theoretical functionals as in the theory of logarithmic Sobolev inequalities: first, the **Kullback information** (or Boltzmann H functional, or Shannon information),

$$H_\mu(\nu) = \int h \log h \, d\mu, \qquad \nu = h\mu;$$

and secondly, the **Fisher information**

$$I_\mu(\nu) = \int \frac{|\nabla h|^2}{h} \, d\mu, \qquad \nu = h\mu.$$

Recall that a probability measure μ on $\mathbb{R}^N$ satisfies a *logarithmic Sobolev inequality* if there is a constant $\lambda > 0$ such that

$$H_\mu(\nu) \leq \frac{1}{2\lambda} I_\mu(\nu),$$

for all probability measures ν on $\mathbb{R}^N$ (with the convention that $H_\mu(\nu) = I_\nu(\mu) = +\infty$ if ν is not absolutely continuous with respect to μ).

As a main difference with respect to the classical theory, I shall *distort* the Fisher information by using a suitable field of quadratic forms; that is, replace $\int |\nabla h|^2/h \, d\mu$ by $\int \langle S\nabla h, \nabla h\rangle/h \, d\mu$, where $x \to S(x)$ is a nonnegative matrix-valued function such that $S(x) \geq \kappa I_N$ for some $\kappa > 0$, independently of x. It turns out that the same algebraic tricks which worked in a Hilbertian context will also work here, at the price of more stringent assumptions on the vector fields: The proofs will be based on some slightly miraculous-looking computations, which may be an indication that there is more structure to understand.

Theorem 28 below is the main result of this section. Note carefully that it is not expressed in terms of linear operators in abstract Hilbert spaces, but in terms of derivation operators on $\mathbb{R}^N$. (The theorem might possibly be generalized by replacing $\mathbb{R}^N$ by a smooth manifold.) So as not to be bothered with regularity issues, I shall assume here that the reference density is rapidly decaying and that all coefficients are C^∞ and have at most polynomial growth; but of course these assumptions can be relaxed. I shall also assume that the solution is smooth if the initial datum is smooth.

THEOREM 28. *Let $E \in C^2(\mathbb{R}^N)$, such that e^{-E} is rapidly decreasing, and $\mu(dX) = e^{-E(X)} \, dX$ is a probability measure on $\mathbb{R}^N$. Let $(A_j)_{1\leq j \leq m}$ and B be first-order derivation operators with smooth coefficients. Denote by A_j^* and B^* their respective adjoints in $L^2(\mu)$, and assume that $B^* = -B$. Denote by A the collection $(A_1, \ldots, A_m)$, viewed as an unbounded operator whose range is made of functions valued in $\mathbb{R}^m$. Define*

$$L = A^*A + B = \sum_{j=1}^m A_j^* A_j + B,$$

---

[2]Of course this does not contradict the fact that it will be *locally* C^∞. The most basic illustration is the case of the linear Fokker–Planck equation $\partial_t h = \Delta_v h - v \cdot \nabla_v h$: Hypercontractivity theory tells us that the semigroup at time t is regularizing from L^p to L^q if and only if $t \geq \log((q-1)/(p-1))/2$, which is finite only if $p > 1$.

and assume that e^{-tL} defines a well-behaved semigroup on a suitable space of positive functions (for instance, $e^{-tL}h$ and $\log(e^{-tL}h)$ are C^∞ and all their derivatives grow at most polynomially if h is itself C^∞ with all derivatives bounded, and h is bounded below by a positive constant).

Next assume the existence of an integer $N_c \geq 1$, first-order derivation operators $C_0, \ldots, C_{N_c+1}$, $R_1, \ldots, R_{N_c+1}$; and vector-valued functions $Z_1, \ldots, Z_{N_c+1}$ (all of them with C^∞ coefficients, growing at most polynomially, as their partial derivatives) such that

$$C_0 = A, \qquad [C_j, B] = Z_{j+1} C_{j+1} + R_{j+1} \quad (0 \leq j \leq N_c), \qquad C_{N_c+1} = 0,$$

and

(i) $[A, C_k]$ is pointwise bounded relatively to A;
(ii) $[C_k, A^*]$ is pointwise bounded relatively to $I, \{C_j\}_{0 \leq j \leq k}$;
(iii) R_k is pointwise bounded with respect to $\{C_j\}_{0 \leq j \leq k-1}$;
(iv) there are positive constants λ_j, Λ_j such that $\lambda_j \leq Z_j \leq \Lambda_j$;
(v) $[A, C_k]^*$ is pointwise bounded relatively to I, A.

Then there is a function $x \to S(x)$, valued in the space of nonnegative symmetric $N \times N$ matrices, uniformly bounded, such that if one defines

$$\mathcal{E}(h) := \int h \log h \, d\mu + \int \frac{\langle S \nabla h, \nabla h \rangle}{h} \, d\mu,$$

one has the estimate

$$\frac{d}{dt}\mathcal{E}(e^{-tL}h) \leq -\alpha \int \frac{\langle S \nabla h, \nabla h \rangle}{h} \, d\mu,$$

for some positive constant $\alpha > 0$, which is explicitly computable in terms of the bounds appearing implicitly in conditions (i)–(v).

If furthermore
(a) there is a positive constant λ such that $\sum_k C_k^* C_k \geq \lambda I_N$, pointwise on $\mathbb{R}^N$;
(b) μ satisfies a logarithmic Sobolev inequality with constant K;
then $S(x)$ is uniformly positive definite, and there is a constant $\kappa > 0$ such that

$$\frac{d}{dt}\mathcal{E}(e^{-tL}h) \leq -\kappa \, \mathcal{E}(e^{-tL}h).$$

In particular,

$$I_\mu((e^{-tL}h\,\mu) = O(e^{-\kappa t}), \qquad H_\mu((e^{-tL}h\,\mu) = O(e^{-\kappa t}),$$

and all the constants in this estimate can be estimated explicitly in terms of the bounds appearing implicitly in conditions (i)–(v), and the constants λ, K.

REMARK 29. The matrices $S(x)$ will be constructed from the vector fields entering the equation, by linear combinations *with constant coefficients*. For more degenerate situations, it may be useful to use varying coefficients.

REMARK 30. A major difference between the assumptions of Theorem 24 and the assumptions of Theorem 28 is that the latter impose *pointwise* bounds on $\mathbb{R}^N$, in the following sense. First, A is an m-tuple of derivation operators $(A_i)_{1 \leq i \leq m}$, each of which can be identified with a vector field σ_i, in such a way that $A_i f = \sigma_i \cdot \nabla f$; so $\sigma = (\sigma_i)_{1 \leq i \leq m}$ can be seen as a map valued in $(m \times N)$ matrices. Then each commutator C_k is also an m-tuple of derivation operators $(C_{k,j})_{1 \leq j \leq m}$, so that $C_{k,j}$ has been obtained from the commutation of $C_{k-1,j}$ with B. Then $[A_i, C_{k,j}]$ is represented by a vector field $\xi_{i,j,k}$; and Assumption (i) says that $|\xi_{i,j,k}(x)|$ is bounded, for all x, by $c\|\sigma(x)\|$, where c is a constant. The other pointwise conditions are to be interpreted similarly. Let us consider for instance Assumption (ii). Since A_i is a derivation, the adjoint of A_i takes the form $-A_i + a_i$, where a_i is a function; so the adjoint of A is of the form $(g_1, \ldots, g_m) \to -\sum A_i g_i + \sum a_i g_i$. Then the commutator of C with A^* is the same as the commutator of C with A, up to an array of operators which are the multiplication by the *functions* $C_j a_i$. So Assumption (ii) really says that the functions $C_j a_i$ are all bounded. Finally, note that in Assumption (a), each C_k is an m-tuple of derivations, so it can be identified to a function valued in $m \times N$ matrices; and $\sum_k C_k^* C_k$ to a function valued in $N \times N$ matrices, which should be uniformly positive definite.

REMARK 31. Theorem A.18 in Appendix A.21 will show that the Assumptions of Theorem 28 entail an immediate "entropic" regularization effect: If the initial datum is only assumed to have finite entropy, then the functional $\mathcal{E}$ becomes immediately finite. This allows to extend the range of application of the method to initial data with very little regularity. In the case of the Fokker–Planck equation I shall show later how to relax even this assumption of finite entropy.

The key to the proof of Theorem 28 is the following lemma, which says that the computations arising in the time-differentiation of the functional $\mathcal{E}$ are quite the same as the computations arising in Theorem 24, *provided that A and C_k commute*.

In the next statement, I shall use the notation
$$\left(\frac{d}{dt}\right)_S \mathcal{F}(h)$$
for the time-derivative of the functional $\mathcal{F}$ along the semigroup generated by the linear operator $-S$. More explicitly,
$$\left(\frac{d}{dt}\right)_S \mathcal{F}(h) = \left.\frac{d}{dt}\right|_{t=0} \mathcal{F}(e^{-tS} h).$$
Moreover, when no measure is indicated this means that the Lebesgue measure should be used.

LEMMA 32. *Let $\mu(dX) = e^{-E(X)} \, dX$, $A = (A_1, \ldots, A_m)$, B and $L = A^*A + B$ be as in Theorem 28. Let $C = (C_1, \ldots, C_m)$ and $C' = (C'_1, \ldots, C'_m)$ be m-tuples of derivation operators on $\mathbb{R}^N$ (all of them with smooth coefficients whose derivatives grow at most polynomially). Then, with the notation $f = he^{-E}$, $u = \log h$, one has*

(6.1) $$\left(\frac{d}{dt}\right)_B \int h \log h \, d\mu = 0;$$

(6.2) $$-\left(\frac{d}{dt}\right)_{A^*A} \int h \log h \, d\mu = \int \frac{|Ah|^2}{h} \, d\mu = \int f |Au|^2,$$

where by convention $|Au|^2 = \sum_i (A_i u)^2$;

(6.3) $$-\left(\frac{d}{dt}\right)_B \int \frac{\langle Ch, C'h \rangle}{h} \, d\mu = \int \frac{\langle Ch, [C', B]h \rangle}{h} \, d\mu + \int \frac{\langle [C, B]h, C'h \rangle}{h} \, d\mu$$

(6.4) $$= \int f \langle Cu, [C', B]u \rangle + \int f \langle [C, B]u, C'u \rangle,$$

where by convention $\langle [C,B]u, C'u\rangle = \sum_j ([C_j,B]u)(C'_j u)$;

(6.5)
$$-\left(\frac{d}{dt}\right)_{A^*A} \int \frac{\langle Ch, C'h\rangle}{h}\, d\mu = 2\int f\langle CAu, C'Au\rangle$$
$$+ \left(\int f\,\langle [C, A^*]Au, C'u\rangle + \int f\,\langle Cu, [C', A^*]Au\rangle\right)$$
$$+ \left(\int f\,\langle CAu, [A, C']u\rangle + \int f\,\langle [A,C]u, C'Au\rangle\right)$$
$$+ \int f\, Q_{A,C,C'}(u),$$

where by convention $\langle Cu, [C', A^*]Au\rangle = \sum_{ij}(C_j u)([C'_j, A_i^*]A_i u)$, etc. and

(6.6) $Q_{A,C,C'}(u) := [A,C]^*(Au \otimes C'u) + [A,C']^*(Au \otimes Cu)$
$$:= \sum_{ij}[A_i, C_j]^*(A_i u\, C'_j u) + [A_i, C'_j]^*(A_i u\, C_j u).$$

REMARK 33. If $[A_i, C_j] = [A_i, C'_j] = 0$ for all i, j, then obviously $Q_{A,C,C'}$ vanishes identically. The same conclusion holds true if $A = C = C'$, even if $[A_i, A_j]$ is not necessarily 0. Indeed, $[A_j, A_i]^*(A_j u\, A_i u) = -[A_i, A_j]^*(A_i u\, A_j u)$, so $\sum_{ij}[A_i, A_j]^*(A_i u\, A_j u) = 0$ by the symmetry $i \leftrightarrow j$. I don't know whether there are simple general conditions for the vanishing of $Q_{A,C,C'}$, that would encompass both $[A, C'] = [A, C] = 0$ and $A = C = C'$ as particular cases.

REMARK 34. One of the conclusions of this lemma is that the time-derivatives of the quantities $\int h \log h\, d\mu$ and $\int \langle Ch, C'h\rangle/h\, d\mu$ can be computed just as the time-derivatives of the quantities $\int h^2\, d\mu$ and $\int \langle Ch, C'h\rangle\, d\mu$, if one replaces in the final result the measure μ by $f = h\mu$, and in the integrand the function h by its logarithm, as long as the quantity $Q_{A,C,C'}$ vanishes. In the special case when $C = C' = A$, this principle is well-known in the theory of logarithmic Sobolev inequalities, where it is stated in terms of Bakry and Émery's "Γ_2 calculus". As in the theory of Γ_2 calculus, Ricci curvature should play a crucial role here, since it is related to the commutator $[A, A^*]$. (In a context of Riemannian geometry, this is what Bochner's formula is about.)

PROOF OF LEMMA 32. The proofs of (6.1) and (6.2) are easy and well-known, however I shall recall them for completeness. The proof of (6.3) will not cause any difficulty. But the proof of (6.5) will be surprisingly complicated and indirect, which might be a indication that a more appropriate formalism is still to be found.

As before, I shall assume that the function h is very smooth, and that all the integrations by parts or other manipulations needed in the proof are well justified. To alleviate notation, I shall abbreviate $e^{-tL}h$ into just h, the time dependence being implicit. Also $f = he^{-E}$ and $u = \log h = \log f + E$ will depend implicitly on the time t. Recall that the Lebesgue measure is used if no integration measure is specified.

By the chain-rule,

$$-\left(\frac{d}{dt}\right)_B \int h \log h \, d\mu = \int (\log h + 1)(Bh) \, d\mu = \int B(h \log h) \, d\mu$$

$$= \int (B^*1)(h \log h) \, d\mu,$$

and this quantity vanishes since $B^* = -B$ is a derivation. This proves (6.1).

Next,

$$-\left(\frac{d}{dt}\right)_{A^*A} \int h \log h \, d\mu = \int (\log h + 1)(A^*Ah) \, d\mu$$

$$= \int \langle A(\log h + 1), Ah \rangle \, d\mu = \int \langle \frac{Ah}{h}, Ah \rangle \, d\mu.$$

This proves (6.2).

To prove (6.3), it suffices to remark that (a) the integrand can be written as a quadratic expression of $\sqrt{h}$ since, by chain rule,

$$\int \frac{\langle Ch, C'h \rangle}{h} \, d\mu = 4 \int \langle C\sqrt{h}, C'\sqrt{h} \rangle \, d\mu;$$

and that (b) the evolution equation for $\sqrt{h}$ along B is the same as for h: indeed, $\partial_t h + Bh = 0$ implies $\partial_t \sqrt{h} + B\sqrt{h} = 0$. So to compute the time-derivative in (6.3) it all reduces to a *quadratic* expression:

$$\int \langle CB\sqrt{h}, C'\sqrt{h} \rangle \, d\mu = \int \langle BC\sqrt{h}, C'\sqrt{h} \rangle \, d\mu + \int \langle [C,B]\sqrt{h}, C'\sqrt{h} \rangle \, d\mu.$$

Then the first term in the right-hand side vanishes since B is antisymmetric. Formula (6.3) follows upon use of the chain-rule again.

Now it only remains to establish (6.5). Before starting the computations, let us recast the equation $\partial_t h + A^*Ah = 0$ in terms of $f = he^{-E}$. It follows by Proposition 5, with $\rho_\infty = e^{-E}$ and $B = 0$, that

$$\partial_t f = \nabla \cdot (D(\nabla f + f \nabla E)) = \nabla \cdot (Df \nabla u),$$

with the diffusion matrix $D = A^*A$, or more rigorously $\sigma^*\sigma$, where σ is such that $Ah = \sigma(\nabla h)$. In particular, if v and w are two smooth functions, then

(6.7) $$\langle D\nabla v, \nabla w \rangle = \langle Av, Aw \rangle.$$

Another relation will be useful later: by explicit computation, if g is a vector-valued smooth function, then

$$A^*g = \nabla \cdot (\sigma^*g) - \langle \sigma \nabla E, g \rangle;$$

it follows that, for any real-valued smooth function u,

(6.8) $$\nabla \cdot (D\nabla u) - \langle D\nabla E, \nabla u \rangle = A^*Au.$$

Next, by chain-rule,

$$\int \frac{\langle Ch, C'h \rangle}{h} \, d\mu = \int f \langle Cu, C'u \rangle.$$

So the left-hand side of (6.5) is equal to

$$(6.9) \quad -\int \nabla \cdot (fD\nabla u)\langle Cu, C'u\rangle - \int f\left\langle C\left(\frac{\nabla \cdot (Df\nabla u)}{f}\right), C'u\right\rangle$$
$$-\int f\left\langle Cu, C'\left(\frac{\nabla \cdot (Df\nabla u)}{f}\right)\right\rangle.$$

The three terms appearing in the right-hand side of (6.9) will be considered separately. First, by integration by parts and (6.7),

$$(6.10) \quad -\int \nabla \cdot (fD\nabla u)\langle Cu, C'u\rangle = \int f\left\langle D\nabla u, \nabla\langle Cu, C'u\rangle_{\mathbb{R}^m}\right\rangle_{\mathbb{R}^N}$$

$$(6.11) \qquad\qquad\qquad\qquad\qquad = \int f\left\langle Au, A\langle Cu, C'u\rangle\right\rangle.$$

For the second term in (6.9), we use the identity

$$\nabla \cdot (Df\nabla u) = f\nabla \cdot (D\nabla u) + \langle D\nabla u, \nabla f\rangle = f\nabla \cdot (D\nabla u) + f\langle D\nabla u, \nabla \log f\rangle.$$

So

$$-\int f\left\langle C\left(\frac{\nabla \cdot (Df\nabla u)}{f}\right), C'u\right\rangle = -\int f\left\langle C\nabla \cdot (D\nabla u), C'u\right\rangle$$
$$-\int f\left\langle C\langle D\nabla u, \nabla \log f\rangle_{\mathbb{R}^N}, C'u\right\rangle_{\mathbb{R}^m}$$
$$= -\int f\left\langle C\nabla \cdot (D\nabla u), C'u\right\rangle - \int f\left\langle C\langle D\nabla u, \nabla u\rangle_{\mathbb{R}^N}, C'u\right\rangle_{\mathbb{R}^m}$$
$$+ \int f\left\langle C\langle D\nabla u, \nabla E\rangle_{\mathbb{R}^N}, C'u\right\rangle_{\mathbb{R}^m}.$$

By combining the first and third integrals in the expression above, then using (6.8) and (6.7) again, we find that

$$-\int f\left\langle C\left(\frac{\nabla \cdot (Df\nabla u)}{f}\right), C'u\right\rangle$$
$$(6.12) \quad = -\int f\left\langle C(\nabla \cdot (D\nabla u) - \langle D\nabla E, \nabla u\rangle_{\mathbb{R}^N}), C'u\right\rangle_{\mathbb{R}^m}$$
$$-\int f\left\langle C\langle D\nabla u, \nabla u\rangle_{\mathbb{R}^N}, C'u\right\rangle_{\mathbb{R}^m}$$
$$= \int f\left\langle CA^*Au, C'u\right\rangle - \int f\left\langle C|Au|^2, C'u\right\rangle$$
$$(6.13) \quad = \int f\left\langle [C, A^*]Au, C'u\right\rangle + \int f\left\langle A^*CAu, C'u\right\rangle$$
$$-2\int f\left\langle (CAu)\cdot(Au), C'u\right\rangle.$$

(In the last term, the dot is just here to indicate the evaluation of the matrix CAu on the vector Au. Also $\langle A^*CAu, C'u\rangle$ should be understood as $\sum_{ij}\langle A_i^*C_jA_iu, C_j'u\rangle$.)

Now the second integral in (6.13) needs some rewriting. By using the chain rule as before, and the definition of the adjoint,

$$(6.14) \quad \int f \langle A^*CAu, C'u \rangle = \int \langle A^*CAu, hC' \log h \rangle \, d\mu$$

$$= \int \langle A^*CAu, C'h \rangle \, d\mu$$

$$= \int \langle CAu, AC'h \rangle \, d\mu$$

$$(6.15) \quad = \int \langle CAu, [A, C']h \rangle \, d\mu + \int \langle CAu, C'Ah \rangle \, d\mu.$$

The first term in (6.15) can be rewritten as

$$(6.16) \quad \int \langle CAu, [A, C']u \rangle \, h \, d\mu = \int f \langle CAu, [A, C']u \rangle.$$

As for the second term in (6.15), since C' is a derivation, it can be recast as

$$\int \langle CAu, C'(hAu) \rangle \, d\mu = \int \langle CAu, hC'Au \rangle \, d\mu + \int \langle CAu, (C'h) \otimes Au \rangle \, d\mu$$

$$(6.17) \quad = \int f \langle CAu, C'Au \rangle + \int f \langle (CAu) \cdot (Au), C'u \rangle.$$

Note that there is a partial simplification with the last term of (6.13) (only partial since the coefficients are not the same).

Of course, the expressions which we obtained for the second term in (6.9) also hold for the third term, up to the exchange of C and C'. After gathering all these results, we find

$$(6.18) \quad -\left(\frac{d}{dt}\right)_{A^*A} \int f \langle Cu, C'u \rangle = \int f \langle Au, A\langle Cu, C'u \rangle \rangle$$

$$(6.19) \quad + \left(\int f \langle [C, A^*]Au, C'u \rangle + \int f \langle Cu, [C', A^*]Au \rangle \right)$$

$$(6.20) \quad + \left(\int f \langle CAu, [A, C']u \rangle + \int f \langle [A, C]u, C'Au \rangle \right)$$

$$(6.21) \quad + 2 \int f \langle CAu, C'Au \rangle$$

$$(6.22) \quad - \left(\int f \langle (CAu) \cdot Au, C'u \rangle + \int f \langle (C'Au) \cdot Cu, Au \rangle \right).$$

The terms appearing in (6.19), (6.20) and (6.21) coincide with some of the ones which appear in (6.5), so it only remains to check that the ones in (6.18) and (6.22) add up to (6.6). By using the identity

$$A\langle Cu, C'u \rangle = (ACu) \cdot (C'u) + (AC'u) \cdot (Cu),$$

we see that the sum of (6.18) and (6.22) can be recast as

$$\int f \left\langle Au, \langle (AC - CA)u, C'u \rangle \right\rangle + \int f \left\langle Au, \langle (AC' - C'A)u, Cu \rangle \right\rangle$$

$$= \int f \left\langle Au, \langle [A, C]u, C'u \rangle \right\rangle + \int f \left\langle Au, \langle [A, C']u, Cu \rangle \right\rangle;$$

or, more explicitly:

$$(6.23) \quad \sum_{ij} \int f(A_i u)([A_i, C_j]u)(C'_j u) + \sum_{ij} \int f(A_i u)([A_i, C'_j]u)(C_j u).$$

It remains to check that (6.23) can be transformed into (6.6). Consider for instance the first term in (6.23), for some index (i,j). Since $[A_i, C_j]$ is a derivation,

$$\int f(A_i u)([A_i, C_j]u)(C'_j u) = \int (A_i u)([A_i, C_j]h)(C'_j u)\, d\mu$$
$$= \int h[A_i, C_j]^*(A_i u\, C'_j u)\, d\mu$$
$$= \int f[A_i, C_j]^*(A_i u\, C'_j u).$$

This concludes the proof of Lemma 32. $\square$

PROOF OF THEOREM 28. Here I shall use the same conventions as in the proof of Lemma 32. The functional $\mathcal{E}$ will be searched for in the form

$$\mathcal{E}(h) = \int fu + \sum_{k=0}^{N_c} \left(a_k \int f|C_k u|^2 + 2b_k \int f\langle C_k u, C_{k+1} u\rangle \right).$$

In other words, the quadratic form S in the statement will be looked for in the form

$$\langle S(x)\xi, \xi\rangle_m = \sum a_k |C_k(x)\xi|^2_{\mathbb{R}^m} + 2\sum b_k \langle C_k(x)\xi, C_{k+1}(x)\xi\rangle_{\mathbb{R}^m},$$

where C_k is identified with a function valued in $m \times N$ matrices.

If the inequalities (5.3) are enforced, then for δ small enough

$$\langle S(x)\xi, \xi\rangle_m \geq K \sum |C_k(x)\xi|^2_m;$$

then S will be a nonnegative symmetric matrix, as desired.

Next, we consider the evolution of $\mathcal{E}$ along the semigroup. As recalled in Lemma 32,

$$-\frac{d}{dt}\int h\log h\, d\mu = +\int \frac{|Ah|^2}{h}\, d\mu.$$

Next,

$$\begin{cases} -\dfrac{d}{dt}\int \dfrac{|C_k h|^2}{h}\, d\mu = 2\left((\mathrm{I})^k_A + (\mathrm{I})^k_B\right), \\[1em] -\dfrac{d}{dt}\int \dfrac{\langle C_k h, C_{k+1} h\rangle}{h}\, d\mu = (\mathrm{II})^k_A + (\mathrm{II})^k_B, \end{cases}$$

where the subscript A indicates the contribution of the A^*A operator, and the subscript B indicates the contribution of the B operator. The goal is to show that these terms can be handled in exactly the same way as in Theorem 24: Everything can be controlled in terms of the quantities

$$(6.24) \quad \int f|C_k Au|^2 \quad \text{and} \quad \int f|C_k u|^2 = \int \frac{|C_k h|^2}{h}\, d\mu.$$

(These integrals play the role that the quantities $\|C_k Ah\|^2$ and $\|C_k h\|^2$ were playing in the proof of Theorem 24.)

6. HYPOCOERCIVITY IN ENTROPIC SENSE

The terms $(I)_B^k$ and $(II)_B^k$ are most easily dealt with. By Lemma 32, we just have to reproduce the result of the computations in the proof of Theorem 24 and divide the integrand by h. So in place of

$$\int \langle C_k h,\, Z_{k+1} C_{k+1} h \rangle \, d\mu + \int \langle C_k h,\, R_{k+1} h \rangle \, d\mu,$$

we have

$$(I)_B^k = \int \frac{\langle C_k h,\, Z_{k+1} C_{k+1} h \rangle}{h} \, d\mu + \int \frac{\langle C_k h,\, R_{k+1} h \rangle}{h} \, d\mu.$$

Then we proceed just as in the proof of Theorem 24: By Cauchy–Schwarz inequality (applied here for vector-valued functions),

$$(I)_B^k \geq -\Lambda_k \sqrt{\int \frac{|C_k h|^2}{h} \, d\mu} \sqrt{\int \frac{|C_{k+1} h|^2}{h} \, d\mu}$$

$$-\sqrt{\int \frac{|C_k h|^2}{h} \, d\mu} \sqrt{\int \frac{|R_{k+1} h|^2}{h} \, d\mu}.$$

Then $|R_{k+1} h|$ can be bounded *pointwise* in terms of $|Ah|, \ldots, |C_k h|$, so $\int |R_{k+1} h|^2 / h \, d\mu$ can be controlled in terms of $\int |C_j h|^2 / h \, d\mu$ for $j \leq k$.

The treatment of $(II)_B^k$ is similar:

$$(II)_B^k = \int \frac{\langle Z_{k+1} C_{k+1} h, C_{k+1} h \rangle}{h} \, d\mu + \int \frac{\langle R_{k+1} h, C_{k+1} h \rangle}{h} \, d\mu$$

$$+ \int \frac{\langle C_k h, Z_{k+2} C_{k+2} h \rangle}{h} \, d\mu + \int \frac{\langle C_k h, R_{k+2} h \rangle}{h} \, d\mu$$

$$\geq \lambda_{k+1} \int \frac{|C_{k+1} h|^2}{h} \, d\mu - \sqrt{\int \frac{|R_{k+1} h|^2}{h} \, d\mu} \sqrt{\int \frac{|C_{k+1} h|^2}{h} \, d\mu}$$

$$- \Lambda_{k+2} \sqrt{\int \frac{|C_k h|^2}{h} \, d\mu} \sqrt{\int \frac{|C_{k+2} h|^2}{h} \, d\mu} - \sqrt{\int \frac{|C_k h|^2}{h} \, d\mu} \sqrt{\int \frac{|R_{k+2} h|^2}{h} \, d\mu}.$$

Then once again, one can control the functions $|R_{k+2} h|$ by $|C_j h|$ for $j \leq k+1$.

Now consider the terms coming from the action of $A^* A$. Let us first pretend that the extra terms $Q_{A,C,C'}$ in (6.5) do not exist. Then by Lemma 32 again,

$$(I)_A^k = \int f |C_k A u|^2 + \int f \langle [C_k, A^*] A u, C_k u \rangle + \int f \langle C_k A u, [A, C_k] u \rangle.$$

By Cauchy–Schwarz inequality (for vector-valued functions),

$$(I)_A^k \geq \int f |C_k A u|^2 - \sqrt{\int f |[C_k, A^*] A u|^2} \sqrt{\int f |C_k u|^2}$$

$$- \sqrt{\int f |C_k A u|^2} \sqrt{\int f |[A, C_k] u|^2}.$$

Then Assumption (iii) implies

$$(6.25) \quad \sqrt{\int f |[C_k, A^*]Au|^2} \leq c\Big(\sqrt{\int f|A^2 u|^2} + \sqrt{\int f|AC_1 u|^2} + \ldots + \sqrt{\int f|AC_k u|^2}\Big).$$

Finally,

$$(II)_A^k = 2\int f \langle C_k Au, C_{k+1} Au\rangle + \int f \langle [C_k, A^*]Au, C_{k+1} u\rangle$$
$$+ \int f \langle C_k u, [C_{k+1}, A^*]Au\rangle + \int f \langle C_k Au, [A, C_{k+1}]u\rangle$$
$$+ \int f \langle [A, C_k]u, C_{k+1} Au\rangle.$$

and this can be bounded below by a negative multiple of

$$\sqrt{\int f|C_k Au|^2}\sqrt{\int f|C_{k+1} Au|^2} + \sqrt{\int f|C_k u|^2}\sqrt{\int f|[C_{k+1}, A^*]Au|^2}$$
$$+ \sqrt{\int f|C_k Au|^2}\sqrt{\int f|[A, C_{k+1}]u|^2} + \sqrt{\int |[A, C_k]u|^2}\sqrt{\int f|C_{k+1} Au|^2};$$

then one can apply (6.25) (as it is, and also with k replaced by $k+1$) to control the various terms above.

All in all, everything can be bounded in terms of the integrals appearing in (6.24), and the computations are *exactly* the same as in the proof of Theorem 24; then the same bounds as in Theorem 24 will work, provided that the coefficients a_k and b_k are well chosen. The result is

$$(6.26) \quad \frac{d}{dt}\mathcal{E}(h) \leq -K \int \frac{\langle S(x)\nabla h(x), \nabla h(x)\rangle}{h(x)} \, d\mu(x).$$

Now let us see what happens if Assumptions (a) and (b) are enforced. By assumption (a), we have $\sum |C_k(x)\xi|^2 \geq \lambda |\xi|^2$, where $\lambda > 0$; so there exists $\kappa > 0$ such that

$$\mathcal{E}(h) \geq \int fu + \kappa \int f|\nabla u|^2 = \int h \log h \, d\mu + \kappa \int \frac{|\nabla h|^2}{h} \, d\mu.$$

Thus $\mathcal{E}$ will dominate both the Kullback information $H_\mu(h\mu)$, and the Fisher information $I_\mu(h\mu)$.

Then, since S is uniformly positive definite,

$$\int \frac{\langle S(x)\nabla h(x), \nabla h(x)\rangle}{h(x)} \, d\mu(x) \geq \lambda \int \frac{|\nabla h|^2}{h} \, d\mu = \lambda \, I_\mu(h\mu).$$

As a consequence, by Assumption (b),

$$\int \frac{\langle S(x)\nabla h(x), \nabla h(x)\rangle}{h(x)} \, d\mu(x) \geq \kappa \, H_\mu(h\mu) = \kappa \int fu$$

for some $\kappa > 0$. So the right-hand side of (6.26) controls also $\int fu$, and in fact there is a positive constant κ such that
$$\frac{d}{dt}\mathcal{E}(h) \leq -\kappa\,\mathcal{E}(h).$$
Then we can apply Gronwall's inequality to conclude the proof of Theorem 28.

It remains to take into account the additional terms generated by $Q_{A,C,C'}$ in (6.5). Precisely, in $(\mathrm{I})_A^k$ we should consider $\int f\, Q_{A,C_k,C_k}(u)$; and in $(\mathrm{II})_B^k$ we should handle $\int f\, Q_{A,C_k,C_{k+1}}(u)$. So the problem is to bound also these expressions in terms of the quantities (6.24).

We start with the additional term in $(\mathrm{I})_A^k$, that is,

(6.27) $$\int f\, Q_{A,C_k,C_k}(u) = \int f[A,C_k]^*(Au \otimes C_k u).$$

By assumption $[A,C_k]^*$ is controlled by I and A, so there is a constant c such that

(6.28) $$\left|\int [A,C_k]^*(Au \otimes C_k u)\right| \leq c\left(\int f\,|A(Au \otimes C_k u)| + \int f\,|Au \otimes C_k u|\right).$$

Next, by the rules of derivation of products,
$$A(AuC_k u) = (A^2 u)(C_k u) + (Au)(AC_k u)$$
$$= (A^2 u)(C_k u) + (Au)(C_k Au) + (Au)([A,C_k]u).$$

Here as in the sequel, I have omitted indices for simplicity; the above equation should be understood as $A_\ell(A_i u\, C_{k,j} u) = (A_\ell A_i u)(C_{k,j} u) + (A_\ell u)(C_{k,j} A_i u) + (A_\ell u)([A_i, C_{k,j}]u)$. Since by assumption $[A,C_k]$ is controlled by $\{C_j\}_{0 \leq j \leq k}$, there exists some constant c such that the following pointwise bounds holds:

$$|A(Au \otimes C_k u)| \leq c\left(|A^2 u|\,|C_k u| + |Au|\,|C_k Au| + |Au|^2 + \sum_{0 \leq j \leq k} |Au|\,|C_j u|\right).$$

Plugging this in (6.28) and then in (6.27), then using the Cauchy–Schwarz inequality, we end up with

$$\left|\int f\, Q_{A,C_k,C_k}(u)\right| \leq c\left(\sqrt{\int f\,|A^2 u|^2}\sqrt{\int f|C_k u|^2}\right.$$
$$+ \sqrt{\int f\,|Au|^2}\sqrt{\int f|C_k Au|^2}$$
$$\left.+ \sum_{j \leq k}\sqrt{\int f\,|Au|^2}\sqrt{\int f|C_j u|^2} + \sqrt{\int f\,|Au|^2}\sqrt{\int f|C_k u|^2}\right).$$

All these terms appear in $(\mathrm{I})_A^k$ with a multiplicative coefficient a_k. So they can be controlled in terms of (6.24), with the right coefficients, as in the proof of Theorem 24, if

$$a_k \ll \max\left(\sqrt{a_0\, b_{k-1}},\ \sqrt{a_k},\ 1,\ \max_{j \leq k}\sqrt{b_{j-1}},\ \sqrt{b_{k-1}}\right).$$

These conditions are enforced by the construction of the coefficients (a_j) and (b_j).

Now we proceed similarly for the additional terms in $(\text{II})_B^k$. By repeating the same calculations as above, we find

$$\left|\int f\, Q_{A,C_k,C_{k+1}}(u)\right| \leq c \left(\sqrt{\int f |Au|^2} \sqrt{\int f |C_{k+1}u|^2} \right.$$

$$+ \sqrt{\int f |A^2 u|^2} \sqrt{\int f |C_{k+1} u|^2}$$

$$+ \sqrt{\int f |Au|^2} \sqrt{\int f |C_{k+1} Au|^2} + \sum_{j \leq k+1} \sqrt{\int f |Au|^2}$$

$$+ \sqrt{\int f |C_j u|^2} + + \sqrt{\int f |Au|^2} \sqrt{\int f |C_k u|^2} + \sqrt{\int f |A^2 u|^2} \sqrt{\int f |C_k u|^2}$$

$$\left. + \sqrt{\int f |Au|^2} \sqrt{\int f |C_k Au|^2} \right)$$

All these terms come with a coefficient b_k, and they are properly controlled by (6.24) if

$$b_k \ll \max(\sqrt{b_k},\ \sqrt{a_0 b_k},\ \max_{j \leq k+1} \sqrt{b_{j-1}},\ \sqrt{a_1}).$$

Again, these estimates are enforced by construction. This concludes the proof of Theorem 28. □

7. Application: the kinetic Fokker–Planck equation

In this section I shall apply the preceding results to the kinetic linear Fokker–Planck equation, which motivated and inspired the proof of Theorem 18 as well as previous works [**14, 34, 32**].

The equation to be studied is (2.6), which I recast here:

(7.1) $$\frac{\partial h}{\partial t} + v \cdot \nabla_x h - \nabla V(x) \cdot \nabla_v h = \Delta_v h - v \cdot \nabla_v h;$$

and the equilibrium measure takes the form

$$\mu(dx\, dv) = \gamma(v)\, e^{-V(x)}\, dv\, dx, \qquad \gamma(v) = \frac{e^{-\frac{|v|^2}{2}}}{(2\pi)^{n/2}}, \qquad \int e^{-V} = 1.$$

Let $\mathcal{H} := L^2(\mu)$, $A := \nabla_v$, $B := v \cdot \nabla_x - \nabla V(x) \cdot \nabla_v$. Then (7.1) takes the form $\partial h/\partial t + Lh = 0$, with $L = A^*A + B$, $B^* = -B$. The kernel $\mathcal{K}$ of L is made of constant functions, and the space $\mathcal{H}^1 = H^1(\mu)$ is the usual L^2-Sobolev space of order 1, with derivatives in both x and v variables, and reference weight μ:

$$\|h\|_{\mathcal{H}^1}^2 = \int_{\mathbb{R}^n \times \mathbb{R}^n} \left(|\nabla_v h(x,v)|^2 + |\nabla_x h(x,v)|^2 \right) \mu(dx\, dv).$$

By direct computation,

$$[A, A^*] = I, \qquad C := [A, B] = \nabla_x, \qquad [A, C] = [A^*, C] = 0,$$
$$[B, C] = \nabla^2 V(x) \cdot \nabla_v.$$

7.1. Convergence to equilibrium in H^1.
In the present case, assumptions (i)–(iii) of Theorem 18 are satisfied if

(7.2) $\nabla^2 V$ is relatively bounded by $\{I, \nabla_x\}$ in $L^2(e^{-V})$.

By Lemma A.24 in Appendix A.23, this is true as soon as there exists a constant $c \geq 0$ such that

(7.3) $$|\nabla^2 V| \leq c(1 + |\nabla V|).$$

The other thing that we should check is the coercivity of $A^*A + C^*C$, which amounts to the validity of the Poincaré inequality of the form

(7.4) $$\int (|\nabla_v h|^2 + |\nabla_x h|^2)\, d\mu \geq \kappa \left[\int h^2\, d\mu - \left(\int h\, d\mu \right)^2 \right].$$

Since μ is the tensor product of a Gaussian distribution in $\mathbb{R}_v^n$ (for which the Poincaré inequality holds true with constant 1) and of the distribution e^{-V} in $\mathbb{R}_x^n$, the validity of (7.4) is equivalent to the validity of a Poincaré inequality (in $\mathbb{R}_x^n$)

(7.5) $$\int |\nabla_x h(x)|^2\, e^{-V(x)}\, dx \geq \lambda \left[\int h^2 e^{-V} - \left(\int h e^{-V} \right)^2 \right].$$

This functional inequality has been studied by many many authors, and it is natural to take it as an assumption in itself. Roughly speaking, inequality (7.5) needs V to grow "at least linearly" at infinity. In Theorem A.1 in Appendix A.19 I recall a rather general sufficient condition for (7.5) to be satisfied; it holds true for instance if (7.3) is true and $|\nabla V(x)| \to \infty$ at infinity. Then Theorem 18 leads to the following statement:

THEOREM 35. *Let V be a C^2 potential in $\mathbb{R}^n$, satisfying conditions (7.3) and (7.5). Then, with the above notation, there are constants $C \geq 0$ and $\lambda > 0$, explicitly computable, such that for all $h_0 \in H^1(\mu)$,*

$$\left\| e^{-tL} h_0 - \int h_0\, d\mu \right\|_{H^1(\mu)} \leq C e^{-\lambda t} \|h_0\|_{H^1(\mu)}.$$

REMARK 36. Conditions (7.3) and (7.5) morally mean that the potential V should grow at least linearly, and at most exponentially fast at infinity. These conditions are more general than those imposed by Helffer and Nier [32] [3] in that no regularity at order higher than 2 is needed, and there is no restriction of polynomial growth on V. Here is a more precise comparison: Helffer and Nier prove exponential convergence under two sets of assumptions: on one hand, [32, Assumption 5.6]; on the other hand, [32, Assumption 5.7] plus a spectral gap condition which is equivalent to (7.5). Both these assumptions 5.6 and 5.7 contain [32, eq.(5.17)], which is stronger than (7.3). Finally, the spectral gap condition is not made explicitly in [32, Assumption 5.6], but is actually a consequence of that assumption, since it implies (A.19.1).

PROOF OF THEOREM 35. We already checked all the assumptions of Theorem 18, except for the existence of a convenient dense subspace $\mathcal{S}$. If V is C^∞, it is possible to choose the space of all C^∞ functions on $\mathbb{R}_x^n \times \mathbb{R}_v^n$ whose derivatives of all

---

[3] This comparison should not hide the fact that the estimates by Helffer and Nier were already remarkably general, and constituted a motivation for the genesis of this paper.

orders vanish at infinity faster than any inverse power of $(1+|\nabla V|)(1+|v|)$. (Note that the operators appearing in the theorem preserve this space because $|\nabla^2 V|$ is bounded by a multiple of $1+|\nabla V|$.) Then there only remains the problem of approximating V by a C^∞ potential, without damaging Condition (7.3). This can be done by a standard convolution argument: let $V_\varepsilon := V * \rho_\varepsilon$, where $\rho_\varepsilon(x) = \varepsilon^{-n}\rho(x/\varepsilon)$, and ρ is C^∞, supported in the unit ball, nonnegative and of unit integral. Then, for all $\varepsilon > 0$,

$$|\nabla^2 V_\varepsilon(x)| \leq \sup_{|x-y|\leq \varepsilon} |\nabla^2 V(y)| \leq C \sup_{|x-y|\leq \varepsilon} (1+|\nabla V(y)|).$$

But (7.3) implies that $\log(1+|\nabla V|^2)$ is L-Lipschitz ($L = 2C$), so

$$|x-y| \leq \varepsilon \implies \frac{1+|\nabla V(x)|^2}{1+|\nabla V(y)|^2} \leq e^{L\varepsilon}.$$

In particular, by (7.3), $|\nabla^2 V(y)|$ can be controlled in terms of $|\nabla V(x)|$, for y close to x. It follows

$$|x-y| \leq \varepsilon \implies |\nabla V(x) - \nabla V(y)| \leq C\left(1+|\nabla V(x)|\right)\varepsilon e^{L\varepsilon}.$$

As a consequence,

$$|\nabla V_\varepsilon(x) - \nabla V(x)| \leq C\left(1+|\nabla V(x)|\right)\varepsilon e^{L\varepsilon}.$$

From this one easily deduces $1+|\nabla V_\varepsilon(x)| \geq (1-C'\varepsilon)(1+|\nabla V(x)|)$, for some explicit constant C'. Then, V_ε satisfies the same condition (7.3) as V, up to the replacement of the constant C by some constant $\widetilde{C}(\varepsilon)$ which converges to C as $\varepsilon \to 0$.

All in all, after replacing V by V_ε, we can apply the first part of Theorem 18 to get

$$(7.6) \quad ((h_\varepsilon(t), h_\varepsilon(t))) + K \int_0^t \left(\int (|\nabla_x h_\varepsilon(s)|^2 + |\nabla_v h_\varepsilon(s)|^2)e^{-V_\varepsilon(x)}\gamma(v)\right) ds$$
$$\leq ((h_0, h_0)),$$

where $h_\varepsilon(t) = e^{-tL_\varepsilon}(h_0 - \int h_0)$, and K is a constant independent of ε.

By the uniqueness theorem of Appendix A.20, $h_\varepsilon(t)$ converges to $h(t) = e^{-tL}h$, in distributional sense as $\varepsilon \to 0$. Also, $\int h_0 e^{-V_\varepsilon}\gamma \longrightarrow \int h_0 e^{-V}\gamma$. Since the left-hand side is a convex functional of h and V_ε converges locally uniformly to V, inequality (7.6) passes to the limit as $\varepsilon \to 0$. The Poincaré inequality for e^{-V} and the definition of the auxiliary scalar product guarantee the existence of $K' > 0$ such that

$$((h(t), h(t))) + K'\int_0^t ((h(s), h(s))) \, ds \leq ((h_0, h_0)).$$

The exponential convergence of $((h(t), h(t)))$ to 0 follows, and the theorem is proved. □

7.2. Explicit estimates. As a crude test of the effectiveness of the method, one can repeat the proof of Theorem 18 on the particular example of the Fokker–Planck equation, taking advantage of the extra structure to get more precise results.

Using $[A, A^*] = I$, one obtains

(7.7)
$$((h, Lh)) \geq \|Ah\|^2 + a\|A^2h\|^2 + b\|Ch\|^2 + c\|CAh\|^2 - (E),$$
$$(E) := a(\|Ah\|^2 + \|Ah\|\,\|Ch\|)$$
$$+ b(\|Ah\|\,\|R_2h\| + 2\|A^2h\|\,\|CAh\| + \|Ah\|\,\|Ch\|)$$
$$+ c\|Ch\|\,\|R_2h\|.$$

Moreover, $R_2 h = -[B, C] = \nabla^2 V \cdot A$; to simplify computations even more, assume that $|\nabla^2 V| \leq M$ (in Hilbert-Schmidt norm, pointwise on $\mathbb{R}^n$). Then

$$(E) \leq a(\|Ah\|^2 + \|Ah\|\,\|Ch\|)$$
$$+ b\Big(M\|Ah\|^2 + 2\|A^2h\|\,\|CAh\| + \|Ah\|\,\|Ch\|\Big) + cM\|Ah\|\,\|Ch\|$$
$$= (a + bM)\|Ah\|^2 + (a + b + cM)\|Ah\|\,\|Ch\| + 2b\|A^2h\|\,\|CAh\|$$
$$\leq (a + bM + 1/4)\|Ah\|^2 + (a + b + cM)^2\|Ch\|^2 + a\|A^2h\|^2$$
$$+ \frac{b^2}{a}\|CAh\|^2.$$

Since $b^2/a \leq c$, the last two terms above are bounded by the terms in $\|A^2h\|^2$ and $\|CAh\|^2$ in (7.7); so

$$((h, Lh)) \geq [1 - (a + bM + 1/4)]\|Ah\|^2 + [b - (a + b + cM)^2]\|Ch\|^2$$

On the other hand, taking into account the spectral gap assumption on $A^*A + C^*C$,

$$((h, h)) \leq (2a + \kappa^{-1})\|Ah\|^2 + (2c + \kappa^{-1})\|Ch\|^2.$$

So the proof yields a convergence to equilibrium in H^1 like $O(e^{-\overline{\lambda}t})$, where

$$\overline{\lambda} := \sup_{(a,b,c)} \min\left(\frac{1 - (a + bM + \frac{1}{4})}{2a + \kappa^{-1}}, \frac{b - (a + b + cM)^2}{2c + \kappa^{-1}}\right),$$

and the supremum is taken over all triples (a, b, c) with $b^2 \leq ac$.

In the particular (quadratic) case where $\nabla^2 V$ is the identity, one has $M = 1$, $\kappa = 1$; then the choice $a = b = c = 0.05$ yields $\overline{\lambda} = 0.025$, which is off the true (computable) rate of convergence to equilibrium $\lambda = 1/2$ (see [**47**, p. 238–239]) by a factor 20. Thus, even if the method is not extremely sharp, it does yield quite decent estimates.[4] Note that the coefficients a, b, c chosen in the end do not satisfy $c \ll b \ll a$!

7.3. Convergence in L^2. Theorem 35 is stated for H^1 initial data. However, it can be combined with an independent regularity study: Under condition (7.3), one can show that solutions of (7.1) satisfy the estimate

(7.8)
$$0 \leq t \leq 1 \implies \|f(t, \cdot)\|_{H^1(\mu)} \leq \frac{C}{t^{3/2}}\|f(0, \cdot)\|_{L^2(\mu)}.$$

A proof is provided in Appendix A.21. Combined with Theorem 35, this estimate trivially leads to the following statement:

---

[4] As a comparison, the bounds by Hérau and Nier [**34**, formula (4)] yield a lower bound on λ which is around 10^{-4}.

THEOREM 37. *Let V be a C^2 potential in $\mathbb{R}^n$, satisfying conditions (7.3) and (7.5). Then, with the above notation, there are constants $C \geq 0$ and $\lambda > 0$, explicitly computable, such that for all $h_0 \in L^2(\mu)$,*

$$t \geq 1 \Longrightarrow \left\| e^{-tL} h_0 - \int h_0 \, d\mu \right\|_{H^1(\mu)} \leq C e^{-\lambda t} \|h_0\|_{L^2(\mu)}.$$

REMARK 38. The following more precise estimate displays at the same time the convergence to equilibrium and the regularization process: There are positive constants C and λ such that for all $t_0 \in (0,1)$, $t \geq t_0$,

$$\left\| e^{-tL} h_0 - \int h_0 \, d\mu \right\|_{H^1(\mu)} \leq C \frac{e^{-\lambda(t-t_0)}}{t_0^{3/2}} \|h_0\|_{L^2(\mu)}.$$

7.4. Convergence for probability densities. Write $e^{-V}(x)\gamma(v) = e^{-E(x,v)}$, and set $f = e^{-E} h$, then the Fokker–Planck equation (7.1) becomes the kinetic equation for the density of particles:

(7.9) $$\frac{\partial f}{\partial t} + v \cdot \nabla_x f - \nabla V(x) \cdot \nabla_v h = \nabla_v \cdot (\nabla_v f + f v).$$

The previous results show that there is exponential convergence to equilibrium as soon as

$$\int f^2 e^E \, dx \, dv < +\infty.$$

As an integrability estimate, this assumption is not very natural for a probability density; as a decay estimate at infinity, it is extremely strong. The goal now is to establish convergence to equilibrium under much less stringent assumptions on the initial data, maybe at the price of stronger assumptions on the potential V.

An obvious approach to this problem consists in using stronger hypoelliptic regularization theorems. For instance, it was shown by Hérau and Nier [34] that if the initial datum in (7.1) takes the form is only assumed to be a tempered distribution, then the solution at later times lies in $L^2(\mu)$, and in fact takes the form $\sqrt{e^{-E}} g$, where g is C^∞ with rapid decay. Similar results can also be shown by variants of the method exposed in Appendix A.21; for instance one may show that if the initial datum belongs to a negative L^2-Sobolev space of order k then for positive times the solution belongs to a positive L^2-Sobolev space of order k', whatever k and k'. In particular, this approach works fine if the initial datum for (7.9) is a probability measure f_0 satisfying

$$\int e^{E(x,v)/2} f_0(dx\, dv) < +\infty.$$

However this still does not tell anything if we assume only polynomial moment bounds on f_0.

In the next result (apparently the first of its kind), this problem will be solved with the help of Theorem 28, that is, by using an entropy approach.

THEOREM 39. *Assume that*
(a) *$V \in C^2(\mathbb{R}^n)$ with $|\nabla^2 V(x)| \leq C$ for all $x \in \mathbb{R}^n$;*
(b) *the reference measure μ satisfies a logarithmic Sobolev inequality;*
(c) *$f_0(dx\, dv)$ is a probability measure with finite moments of order 2:*

$$\int_{\mathbb{R}^n \times \mathbb{R}^n} (|x|^2 + |v|^2) \, f_0(dx\, dv) < +\infty.$$

Then the solution to (7.9) with initial datum f_0 converges to e^{-E} exponentially fast as $t \to \infty$, in the sense of relative entropy:

$$(7.10) \quad \int f(t,x,v) \log\left(\frac{f(t,x,v)}{e^{-E(x,v)}}\right) dx\, dv = O(e^{-\alpha t}) \qquad (t \geq 1),$$

with explicit estimates.

REMARK 40. A well-known sufficient condition for e^{-V} to satisfy a logarithmic Sobolev inequality is $V = W + w$, where $\nabla^2 W \geq \kappa I_n$, $\kappa > 0$, and w is bounded (this is the so-called "uniformly convex + bounded" setting).

REMARK 41. Here are more precise results in the spirit of Remark 38. Write $H_\mu(h) = \int h \log h\, d\mu$. First, there are positive constants C and λ (only depending on V) such that for all $t_0 \in (0,1)$, $t \geq t_0$,

$$(7.11) \quad H_\mu(e^{-tL}h) \leq \frac{C e^{-\lambda(t-t_0)}}{t_0^3} H_\mu(h).$$

Secondly, one can find an exponent $\beta = \beta(n)$, a constant C depending only on n, V and on $\int (|x|^2 + |v|^2) f_0(dx\, dv)$ such that for all $t_0 \in (0,1)$, $t \geq t_0$,

$$(7.12) \quad H_\mu(e^{-tL}h) \leq \frac{C e^{-\lambda(t-t_0)}}{t_0^\beta}.$$

PROOF OF THEOREM 39. Since ∇V is Lipschitz by assumption, it can be shown by standard techniques that the Fokker–Planck equation admits a unique measure-valued solution. So it is sufficient to establish the convergence for very smooth initial data, with rates that do not depend on the smoothness of the initial datum, and then use a density argument.

I shall give two slightly different proofs of (7.10). The first argument will involve less steps but require more moments and assumptions. The second proof will achieve the generality stated in Theorem 39.

First proof of (7.10): In this proof I shall assume that $V \in C^\infty(\mathbb{R}^n)$ with $|\nabla^j V(x)| \leq C_j$ for all $j \geq 2$ and $x \in \mathbb{R}^n$; and that f_0 has bounded moments of all orders, not just of order 2.

Since $\nabla^2 V$ is bounded, the transport coefficients appearing in (7.9) are Lipschitz (uniformly for $(x,v) \in \mathbb{R}^n_x \times \mathbb{R}^n_v$), and it is easy to show by classical estimates that all moments increase at most linearly in time:

$$\int (1 + |v|^2 + |x|^2)^{k/2} f(t, dx\, dv) = O(1+t).$$

By Theorem A.15 in Appendix A.21, $f(t,\cdot)$ also belongs to all Sobolev spaces (in x and v) for $t > 0$; in fact, estimates of the form

$$\|f(t,\cdot)\|_{H^k_x H^\ell_v(\mathbb{R}^n_x \times \mathbb{R}^n_v)} = O(t^{-\beta(k,\ell)}) \qquad 0 < t \leq 1$$

will be established in that Appendix. Then by elementary interpolation, $f(t,\cdot)$ lies in all weighted Sobolev spaces for all $t \in (0,1)$: That is,

$$\|f(t,\cdot)\|_{H^k_s} := \|f(t,x,v)(1+|v|^2+|x|^2)^{s/2}\|_{H^k} < +\infty.$$

It is shown in [55, Lemma 1] that $I(f) \leq C\|f\|_{H^k_s}$ for k and s large enough (depending on n), where $I(f)$ stands for the Fisher information, $\int f|\nabla(\log f)|^2/f$. So $f(t,\cdot)$ has a finite Fisher information (in both x and v variables) for all $t > 0$.

Since also $f(t, \cdot)$ has all its moments bounded and $|\nabla E| = O(1 + |x| + |v|)$, we have in fact
$$\int f |\nabla (\log f + E)|^2 = O(t^{-\gamma}) \qquad 0 < t \leq 1$$
for some $\gamma > 0$, where the time variable is omitted in the left-hand side. So from time $t = t_0 > 0$ on, the solution f has a finite relative Fisher information with reference measure $\mu(dx\,dv) = e^{-E(x,v)}\,dx\,dv$.

Then we can apply Theorem 28 with $A = \nabla_v$, $B = v \cdot \nabla_x - \nabla V(x) \cdot \nabla_v$, $C_1 = [A, B] = \nabla_x$, $R_1 = 0$, $C_2 = 0$, $R_2 = \nabla^2 V(x) \cdot \nabla_v$, $Z_j = I$. Assumptions (i), (ii), (iii) and (v) in Theorem 28 are automatically satisfied, and Assumption (iv) is also satisfied since $\nabla^2 V$ is bounded. (This is the place where the boundedness of the Hessian of V is crucially used.) Since the reference measure μ is the product of $e^{-V(x)}\,dx$ (which satisfies a logarithmic Sobolev inequality by assumption) with $\gamma(v)\,dv$ (which also satisfies a logarithmic Sobolev inequality), μ itself satisfies a logarithmic Sobolev inequality. So Theorem 28 yields the estimate
$$\int f(\log f + E) = O(e^{-\lambda(t-t_0)})$$
for $t \geq t_0$. In words: The relative entropy of the solution with respect to the equilibrium measure converges to 0 exponentially fast as $t \to \infty$. This concludes the proof of (7.12) and establishes the desired estimate. (To obtain (7.11), replace Theorem A.15 by Theorem A.18.)

Second proof of (7.10): Now I shall work under just the assumptions stated in Theorem 39. As before, the Lipschitz bound on ∇V implies $\int (|x|^2 + |v|^2) f(t, dx\,dv) = O(1+t)$. Then we apply Theorem A.15 in Appendix A.21, but only for $m = 1$ (so we do just as in the proof of Theorem A.8, but with L^1 a priori bounds rather than L^2). For this it is sufficient that $\nabla^2 V$ be bounded. Then we obtain $\int |\nabla_{x,v} f|^2\,dx\,dv = O(t^{-\beta})$ for some $\beta > 0$, so by Nash inequality $\int f^2\,dx\,dv = O(t^{-\gamma})$ for some other $\gamma > 0$. This and the bound on $\int f(|x|^2 + |v|^2)\,dx\,dv$ imply a bound on $\int f \log f$ for any $t > 0$.

Since $\log E = O(1 + |x|^2)$ and f has second-order moments, it follows that $H_\mu(f) = \int f \log(f/E)$ is finite for $t > 0$. At this point we can apply the *entropic hypoelliptic regularization* phenomenon stated in Theorem A.18 and deduce the finiteness of the Fisher information $I_\mu(f)$ for positive time. Then the rest of the argument is based on Theorem 28 as in the first proof above. □

8. The method of multipliers

A crucial ingredient in the L^2 treatment of the Fokker–Planck equation was the use of the mixed second derivative $CAh = \nabla_v \nabla_x h$ to control the error term $[B, C]h = (\nabla^2 V) \cdot \nabla_v h$. There is an alternative strategy, which does not need to use CA: It consists in modifying the quadratic form (4.2) thanks to well-chosen auxiliary operators, typically multipliers. In the case of the Fokker–Planck equation, this method leads to less general results; it is however of independent interest, and can certainly be applied to many equations. In this section I shall present a variant of Theorem 18 allowing for multipliers, and test its applicability to the Fokker–Planck equation. Some extensions are feasible, but I shall not consider them.

Let again A and B be as in Subsection 1.1, and $C = [A, B]$, $R_2 = [C, B]$; assume that $[A, C] = 0$ for simplicity. Let M, N be two self-adjoint, invertible

8. THE METHOD OF MULTIPLIERS

nonnegative operators such that

$$[B, M] = 0, \qquad [B, N] = 0, \qquad [M, N] = 0$$

(these conditions can be somewhat relaxed by imposing only an adequate control on the commutators, but this leads to cumbersome calculations). Instead of (4.2), consider the quadratic form

$$(8.1) \qquad ((h, h)) = \|h\|^2 + a\|MAh\|^2 + 2b\langle MAh, NCh\rangle + c\|NCh\|^2.$$

By straightforward variants of the calculations performed in the proof of Theorem 18, one obtains

$$((h, Lh)) = \|Ah\|^2 + a\|MA^2h\|^2 + b\|\sqrt{MN}Ch\|^2 + c\|NCAh\|^2 - (E),$$

where

$$-(E) := a\Big(\langle MAh, MCh\rangle + \langle MAh, [M, A^*]A^2h\rangle + \langle [A, M]Ah, MA^2h\rangle$$
$$+ \langle MAh, M[A, A^*]Ah\rangle\Big)$$
$$+ b\Big(\langle MAh, N[B, C]h\rangle + 2\langle MA^2h, NCAh\rangle$$
$$+ \langle [A, M]Ah, NCAh\rangle + \langle MAh, [N, A^*]CAh\rangle + \langle [A, N]Ch, MA^2h\rangle$$
$$+ \langle NCh, [M, A^*]A^2h\rangle + \langle NCh, M[A, A^*]Ah\rangle\Big)$$
$$+ c\Big(\langle NCh, NR_2h\rangle + \langle [A, N]Ch, NCAh\rangle + \langle NCh, [N, A^*]CAh\rangle\Big).$$

It would be a mistake to use Cauchy–Schwarz inequality right now. Instead, one should first "re-distribute" the multipliers M and N on the two factors in the scalar products above. For instance, $\langle MAh, MCh\rangle$ is first rewritten $\langle Ah, M^2Ch\rangle$ since it should be controlled by (inter alia) $\|Ah\|^2$, not $\|MAh\|^2$. To obtain the correct weights, one is sometimes led to introduce the inverses M^{-1} and N^{-1}. In the end,

$$(E) \leq a\Big(\|Ah\|\,\|M^2Ch\| + \|Ah\|\,\|M[M, A^*]A^2h\|$$
$$+ \|[A, M]Ah\|\,\|MA^2h\| + \|Ah\|\,\|M^2[A, A^*]Ah\|\Big)$$
$$+ b\Big(\|Ah\|\,\|MN[B, C]h\| + 2\|MA^2h\|\,\|NCAh\| + \|[A, M]Ah\|\,\|NCAh\|$$
$$+ \|Ah\|\,\|M[N, A^*]CAh\| + \|(\sqrt{MN})Ch\|\,\|(\sqrt{N/M})[M, A^*]A^2h\|$$
$$+ \|[A, N]Ch\|\,\|MA^2h\| + \|\sqrt{MN}Ch\|\,\|\sqrt{MN}[A, A^*]Ah\|\Big)$$
$$+ c\Big(+\|(\sqrt{MN})Ch\|\,\|(N^{3/2}/M^{1/2})R_2h\|$$
$$+ \|[A, N]Ch\|\,\|NCAh\| + \|(\sqrt{MN})Ch\|\,\|(\sqrt{N/M})[N, A^*]CAh\|\Big).$$

Of course, in the above $\sqrt{N/M}$ stands for $N^{1/2}M^{-1/2}$, etc.

Repeating the scheme of the proof of Theorem 18, it is easy to see that (E) can be controlled in a satisfactory way as soon as, say (conditions are listed in order of appearance and the notation of Subsection 1.3 is used),

$$M^2 \preccurlyeq \sqrt{MN}, \quad [M, A^*] \preccurlyeq I, \quad [A, M] \preccurlyeq I, \quad M^2[A, A^*] \preccurlyeq I,$$
$$MN[B, C] \preccurlyeq A, \quad [A, M] \preccurlyeq I, \quad M[N, A^*] \preccurlyeq N, \quad [A, N] \preccurlyeq \sqrt{MN},$$
$$(\sqrt{N/M})[M, A^*] \preccurlyeq M, \quad \sqrt{MN}[A, A^*] \preccurlyeq I, \quad (N^{3/2}/M^{1/2})[B, C] \preccurlyeq A,$$
$$[A, N] \preccurlyeq \sqrt{MN}, \quad (\sqrt{N/M})[N, A^*] \preccurlyeq N.$$

For homogeneity reasons it is natural to assume $N = M^3$. Then the above conditions are satisfied if

(8.2) $$M^2[A, A^*] \preccurlyeq I, \quad M^4[B, C] \preccurlyeq A,$$

(8.3) $$[M, A] \preccurlyeq I, \quad [M, A^*] \preccurlyeq I, \quad [M^3, A] \preccurlyeq M^2, \quad [M^3, A^*] \preccurlyeq M^2.$$

If these conditions are satisfied, then one can repeat the scheme of the proof of Theorem 18, with an important difference: instead of $\|Ah\|^2 + \|Ch\|^2$, it is only $\|Ah\|^2 + \|(\sqrt{MN})Ch\|^2 = \|Ah\|^2 + \|M^2Ch\|^2$ which is controlled in the end. This leads to the following theorem.

THEOREM 42. *With the notation of Subsection 1.1, assume that*

$$[C, A] = 0, \quad [C, A^*] = 0,$$

and that there exists an invertible nonnegative self-adjoint bounded operator M on $\mathcal{H}$, commuting with B, such that conditions (8.2) and (8.3) are fulfilled. Define

(8.4) $$((h, h)) = \|h\|^2 + a\|MAh\|^2 - 2b\langle M^2Ah, Ch\rangle + c\|M^3Ch\|^2.$$

Then, there exists $K > 0$, only depending on the bounds appearing implicitly in (8.2) and (8.3), such that

$$\Re((h, Lh)) \geq K(\|Ah\|^2 + \|M^2Ch\|^2).$$

If in addition

(8.5) $$A^*A + C^*M^4C \quad \text{admits a spectral gap } \kappa > 0,$$

then L is hypocoercive on $\mathcal{H}^1/\mathcal{K}$: there exists constants $C \geq 0$ and $\lambda > 0$, explicitly computable, such that

$$\|e^{-tL}\|_{\mathcal{H}^1/\mathcal{K} \to \mathcal{H}^1/\mathcal{K}} \leq Ce^{-\lambda t}.$$

As usual, it might be better in practice to guess the right multipliers and re-do the proof, than to apply Theorem 42 directly. It is also clear that many generalizations can be obtained by combining the method of multipliers with the methods used in Theorem 24. Rather than going into such developments, I shall just show how to apply Theorem 42 on the Fokker–Planck equation with a potential $V \in C^2(\mathbb{R}^n)$. In that case, $[A^*, A] = I$ and $[B, C] = (\nabla^2 V)A$. When $\nabla^2 V$ is bounded and $A^*A + C^*C$ is coercive, there is no need to introduce an auxiliary operator M: the choice $M = I$ is sufficient to provide exponential convergence to equilibrium. But a multiplier might be useful when $\nabla^2 V$ is unbounded. Assume, to fix ideas, that V behaves at infinity like $O(|x|^{2+\alpha})$ for some $\alpha > 0$, and $|\nabla^2 V|$ like $O(|x|^\alpha)$; then it is natural to use an operator M which behaves polynomially, in such a way as to compensate the divergence of V. In the rest of the section, I shall use this strategy to recover the exponential convergence for the kinetic Fokker–Planck equation under assumptions (8.6) and (8.7) below.

8. THE METHOD OF MULTIPLIERS

Let M be the operator of multiplication by $m(x, v)$, where

$$m(x, v) := \frac{1}{\left(V_0 + V(x) + \frac{|v|^2}{2}\right)^{\frac{\alpha}{4(2+\alpha)}}},$$

and V_0 is a constant, large enough that $V_0 + V$ is bounded below by 1. Since $Bm = 0$ and B is a derivation, it is true that B commutes with M. Assume that $V_0 + V$ is bounded below by a multiple of $1 + |x|^{\alpha+2}$; then

$$m^4 \leq \frac{1}{(V_0 + V(x))^{\frac{\alpha}{(2+\alpha)}}} \leq \frac{K}{(1 + |x|)^{\alpha}}$$

for some constant $K > 0$, and then $m^4(\nabla^2 V)$ is bounded, so that $M^4[B, C]$ is relatively bounded by A. Finally, condition (8.3) reduces to

$$|\nabla_v m| \leq Km,$$

which is easy to check. To summarize, conditions (8.2) and (8.3) are fulfilled as soon as there exist constants $C \geq 0$, $K > 0$ and $\alpha > 0$ such that

(8.6) $\quad \forall x \in \mathbb{R}^n, \quad V(x) \geq K|x|^{2+\alpha} - C, \quad |\nabla^2 V(x)| \leq C(1 + |x|^{\alpha}).$

To recover exponential convergence under these assumptions, it remains to check the spectral gap assumption (8.5)! This will be achieved under the following assumption: there exists a potential W, and constants $C \geq 0$, $K > 0$ such that

(8.7) $\quad \forall x \in \mathbb{R}^n, \quad |V(x) - W(x)| \leq C, \quad \nabla^2 W(x) \geq K(1 + |x|)^{-\alpha}.$

From (8.6) there exists $K > 0$ such that

$$m^4(x, v) \geq \left(\frac{K}{(1 + |v|)^{\frac{1}{(2+\alpha)}}}\right) \frac{1}{(1 + |x|)^{\alpha}} =: \Phi(v)\Psi(x).$$

Then

$$\nabla_x^* m^4 \nabla_x \geq \Phi(\nabla_x^* \Psi \nabla_x).$$

Now I claim that $\nabla_x^* \Psi \nabla_x$ (where Ψ is a shorthand for the multiplication by Ψ) is coercive in $L^2(e^{-V} dx)$, or in other words that there exists $K > 0$ such that

$$\int f e^{-V} = 0 \implies \int \Psi |\nabla_x f|^2 e^{-V} \geq K \int f^2 e^{-V};$$

or equivalently, that there is a constant C such that for all $f \in L^2(e^{-V})$,

$$\int [f(x) - f(y)]^2 e^{-V(x)} e^{-V(y)} \, dx \, dy \leq C \int \Psi(x) |\nabla_x f(x)|^2 e^{-V(x)} \, dx.$$

Indeed, with C standing for various positive constants, one can write

$$\int [f(x) - f(y)]^2 e^{-V(x)} e^{-V(y)} \, dx \, dy$$

$$\leq C \int [f(x) - f(y)]^2 e^{-W(x)} e^{-W(y)} \, dx \, dy$$

$$\leq C \int \langle (\nabla^2 W(x))^{-1} \nabla_x f(x), \nabla_x f(x) \rangle e^{-W(x)} \, dx$$

$$\leq C \int (1 + |x|)^{\alpha} |\nabla_x f(x)|^2 e^{-V(x)} \, dx,$$

where the passage from the first to the second line is justified by the **Brascamp–Lieb inequality** [8, Theorem 4.1].

Now it is possible to conclude: the operator $A^*A = \nabla_v^* \nabla_v$ is coercive on $L^2(\gamma)$, γ standing for the Gaussian distribution in the v variable, and the operator $\nabla_x^* \Psi \nabla_x$ is coercive on $L^2(e^{-V})$. Theorem A.2 in Appendix A.19 shows that $A^*A + \Phi \nabla_x^* \Psi \nabla_x$ is coercive on $L^2(\mu)$, where μ is the equilibrium distribution for the Fokker–Planck equation. By monotonicity, $A^*A + C^*M^4C$ also admits a spectral gap; this was the last ingredient needed for Theorem 42 to apply.

9. Further applications and open problems

A very nice application of Theorem 24 was recently done by Capella, Loeschcke and Wachsmuth on the so-called Landau–Lifschitz–Gilbert–Maxwell model arising in micromagnetism. Under certain simplifying assumptions, the linearized version of this model can be written

(9.1)
$$\begin{cases} \partial_t m = J(h - m) \\ \partial_t h = -\nabla \wedge \nabla \wedge h - J(h - m) \\ \nabla \cdot h = -\nabla \cdot m, \end{cases}$$

where $m : \mathbb{R}^3 \to \mathbb{R}^2$ stands for the (perturbation of the) magnetization, and $h : \mathbb{R}^3 \to \mathbb{R}^3$ for the (perturbation of the) magnetic field; moreover, J is the usual symplectic operator $J[x_1, x_2, x_3] = [-x_2, x_1]$. Obviously, the system (9.1) is dissipative but strongly degenerate, since the dissipation term $-\nabla \wedge \nabla \wedge h$ only acts on h, and not even on all components of h. This case turns out to be particularly degenerate since one needs *three* commutators to apply Theorem 24. For further details I refer the reader to the preprint by Capella, Loeschcke and Wachsmuth [10].

Still, many issues remain open in relation to the hypocoercivity of operators of the form $A^*A + B$. I shall describe four of these open problems below.

9.1. Convergence in entropy sense for rapidly increasing potentials.
In the present paper I have derived some first results of exponential convergence to equilibrium for the kinetic Fokker–Planck equation based on an entropy method (Theorem 39). While these results seem to be the first of their kind, they suffer from the restriction of boundedness imposed on the Hessian of the potential. It is not clear how to relax this assumption in order to treat, say, potentials that behave at infinity like $|x|^\beta$, $\beta > 2$. A first possibility would be to try to adapt the method of multipliers, but then we run into two difficulties: (a) Entropic variants of the Brascamp–Lieb inequality do not seem to be true in general, and are known only under certain particular restrictions on the reference measure (see the discussion by Bobkov and Ledoux [6, Proposition 3.4]); (b) It is not clear that there is an entropic analogue of Theorem A.3. Both problems (a) and (b) have their own interest.

Another option would be to try to relax the *local* conditions (i)–(iv) into *global* (integrated) boundedness conditions, so as to have an analogue of Lemma A.24 where the reference measure would be the solution f of the Fokker–Planck equation. This is conceivable only if f satisfies some good a priori estimates for positive times.

9.2. Application to oscillator chains.
One of the motivations for the present study was the hope to revisit the works by Eckmann, Hairer, Rey-Bellet and others on hypoelliptic equations for oscillator chains, modelling heat diffusion [**19, 20, 21, 45, 46**]. So far I have obtained only very partial success in that direction. If we

try to apply Theorem 24 to the model, as it is described e.g. in the last section of [20], we find that the assumptions of Theorem 24 apply as soon as

(a) the "pinning potential" V_1 and the "interaction potential" V_2 have bounded Hessians;

(b) the Hessian of the interaction potential is bounded below by a positive constant;

(c) the second derivatives of the logarithm of the stationary density are bounded;

(d) the stationary measure satisfies a Poincaré inequality.

Let us discuss these conditions. Assumption (a) is a bit too restrictive, since it excludes for instance the quartic double-well potentials which are classically used in that field; but it would still be admissible for a start; and hopefully this restriction could be relaxed later by a clever use of the method of multipliers. (By the way, it is interesting to note that such assumptions are not covered by the results in [20] which need a superquadratic growth at infinity.) Next, Assumption (b) is not so surprising since (as far as I know) it has been imposed by all authors who worked previously on the subject.[5] But it is a completely open problem to derive sufficient conditions for Assumptions (c) and (d), except in the simple case where the two temperatures of the model are equal. This example illustrates an important remark: The range of application of Theorem 24 (and other theorems of the same kind) will be considerably augmented when one has qualitative theorems about the stationary measure for nonsymmetric diffusion processes. For instance,

- Are there simple conditions implying a Poincaré inequality for the stationary measure?

- Can one derive bounds about the Hessian of the logarithm of its density?

The first question was addressed recently in papers by Röckner and Wang (see for instance [48]) in the context of *elliptic* equations, and it looks like a challenging open problem to extend their results to hypoelliptic equations. The second question seems to be completely open; of course it has its intrinsic interest, since very little has been known so far about the stationary measures constructed e.g. in [21].

9.3. The linearized compressible Navier–Stokes system.

An extremely interesting instance of hypocoercive linear *system* is provided by the linearized compressible Navier–Stokes equations for perfect gases. In this example, the noncommutativity does not arise because of derivation along noncommuting vector fields, but because of the noncommutativity of the space where the unknown takes its values.

Obtained by linearizing the nonlinear system of Section 16 around the equilibrium state $(1, 0, 1)$, the linearized compressible Navier–Stokes system reads as follows:

(9.2)
$$\begin{cases} \partial_t \rho + \nabla \cdot u = 0; \\ \partial_t u + \nabla(\rho + \theta) = \mu \Delta u + \mu \left(1 - \frac{2}{N}\right) \nabla(\nabla \cdot u); \\ \partial_t \theta + \frac{2}{N} \nabla \cdot u = \kappa \Delta \theta. \end{cases}$$

---

[5] More generally, as pointed out to me by Hairer, all existing results seem to require that the interaction potential does dominate the pinning potential.

Here N is the dimension, (ρ, u, θ) are *fluctuations* of the density, velocity and temperature respectively, $\mu > 0$ is the viscosity of the fluid and $\kappa > 0$ the heat conductivity. So it is natural to define $\mathcal{H} = L^2(\Omega; \mathbb{R} \times \mathbb{R}^N \times \mathbb{R})$, where $\Omega \subset \mathbb{R}^N$ is the position domain, and the target space $\mathbb{R} \times \mathbb{R}^N \times \mathbb{R}$ is equipped with the Euclidean norm

$$\left\|(\rho, u, \theta)\right\|^2 = \rho^2 + |u|^2 + \frac{N}{2}\theta^2,$$

which is (up to a factor $-1/2$) the quadratic approximation of the usual entropy of compressible fluids.

Let $h = (\rho, u, \theta)$; it turns out that (9.2) can be written in the form $\partial_t h + Lh = 0$, where $L = A^*A + B$, $B^* = -B$, and A, B are quite simple:

(9.3)
$$\begin{cases} Ah = \left(0, \sqrt{2\mu}\{\nabla u\}, \sqrt{\kappa}\nabla\theta\right) \\ Bh = \left(\nabla \cdot u, \nabla\rho + \nabla\theta, \frac{2}{N}\nabla \cdot u\right). \end{cases}$$

Here I have used the notation

$$\{\nabla u\}_{ij} = \left(\frac{1}{2}\left(\frac{\partial u_i}{\partial x_j} + \frac{\partial u_j}{\partial x_i}\right) - \left(\frac{\nabla \cdot u}{N}\right)\delta_{ij}\right)$$

for the traceless symmetrized (infinitesimal) strain tensor of the fluid.

The system (9.2) is degenerate in two ways. First, the diffusion on the velocity variable u does not control all directions: In general it is false that $\int |\{\nabla u\}|^2$ controls the whole of $\int |\nabla u|^2$ (see the discussion in [**16**] for instance: one needs at least an additional control on the divergence). Secondly, there is no diffusion on the density variable ρ. This suggests to consider commutators between $\widetilde{A} : h \to (0, 0, \nabla\theta)$ and B. After some computations one gets (in slightly sketchy notation)

$$[\widetilde{A}, B] = C_1 + R_1; \qquad [C_1, B] = C_2 + R_2;$$
$$C_1 h = \frac{2}{N}\left(0, 0, \nabla\nabla \cdot u\right); \qquad R_1 h = -(0, \nabla^2\theta, 0);$$
$$C_2 h = \frac{2}{N}(0, 0, \nabla\Delta\rho); \qquad R_2 h = \frac{2}{N}(0, -\nabla^2\nabla \cdot u, \nabla\Delta\theta).$$

So the commutator C_1 controls the variations of the divergence of u, while the iterated commutator C_2 controls the variations of the density ρ. However, if we try to apply Theorem 24 in this situation, we immediately run into problems to control the remainder R_2, and need to modify the strategy. This problem is tricky enough to deserve a separate treatment, so I shall not consider it in this memoir.

9.4. A model problem arising in the study of Oseen vortices. All the material in this subsection was taught to me by Gallay. Oseen vortices are certain self-similar solutions to the two-dimensional incompressible Navier–Stokes equation, in vorticity formulation [**25, 26**]. The linear stability analysis of these vortices reduces to the spectral analysis of the operator $S + \alpha B$ in $L^2(\mathbb{R}^2)$, where

(9.4)
$$\begin{cases} S\omega = -\Delta\omega + \frac{|x|^2}{16}\omega - \frac{\omega}{2}, \\ B\omega = \mathrm{BS}[G] \cdot \nabla\omega + 2\,\mathrm{BS}[G^{1/2}\omega] \cdot \nabla G^{1/2}; \end{cases}$$

here BS[ω] is the velocity field reconstructed from the vorticity ω:

$$\mathrm{BS}[\omega](x) = \frac{1}{2\pi} \int_{\mathbb{R}^2} \frac{(x-y)^\perp}{|x-y|^2} \, \omega(y) \, dy,$$

and $v^\perp$ is obtained from v by rotation of angle $\pi/2$; moreover G is a Gaussian distribution: $G(x) = e^{-|x|^2/4}/(4\pi)$; and α is a real parameter.

The spectral study of $S + \alpha B$ turns out to be quite tricky. In the hope of getting a better understanding, one can decompose ω in Fourier series: $\omega = \sum_{n \in \mathbb{Z}} \omega_n(r) e^{in\theta}$, where (r, θ) are standard polar coordinates in $\mathbb{R}^2$. For each n, the operators S and B can be restricted to the vector space generated by $e^{in\theta}$, and can be seen as just operators on a function $\omega = \omega(r)$:

$$\begin{cases} (S_n \omega)(r) = -\partial_r^2 \omega - \left(\dfrac{r}{2} + \dfrac{1}{r}\right) \partial_r \omega - \left(1 - \dfrac{n^2}{r^2}\right) \omega, \\ (B_n \omega)(r) = i\, n\, (\varphi \omega - g \Omega_n); \end{cases}$$

here $g(r) = e^{-r^2/4}/4\pi$, $\varphi(r) = (1 - e^{-r^2/4})/2\pi r^2$, and $\Omega_n(r)$ solves the differential equation

$$-(r\Omega')' + \frac{n^2}{r}\Omega = \frac{r}{2}\omega.$$

The regime $|\alpha| \to \infty$ is of physical interest and has already been the object of numerical investigations by physicists. There are two families of eigenvalues which are imposed by symmetry reasons; but apart from that, it seems that all eigenvalues converge to infinity as $|\alpha| \to \infty$, and for some of them the precise asymptotic rate of divergence $O(|\alpha|^{1/2})$ has been established by numerical evidence. If that is correct, this means that the "perturbation" of the symmetric part S by the antisymmetric, lower-order operator αB is strong enough to send most eigenvalues to infinity as $|\alpha| \to \infty$. Obviously, this is again the manifestation of a hypocoercive phenomenon.

To better understand this stability issue, Gallagher and Gallay suggested the following

MODEL PROBLEM 43. *Identify sufficient conditions on $f : \mathbb{R} \to \mathbb{R}$, so that the real parts of the eigenvalues of*

$$L_\alpha : \omega \longmapsto (-\partial_x^2 \omega + x^2 \omega - \omega) + i\alpha f \omega$$

in $L^2(\mathbb{R})$ go to infinity as $|\alpha| \to \infty$, and estimate this rate.

Here is how Gallagher and Gallay obtained a first partial solution to this problem. Set $\mathcal{H} = L^2(\mathbb{R}; \mathcal{C})$, $A = \partial_x \omega + x\omega$, $B\omega = (i\alpha f)\omega$. Then $C\omega = i\alpha f'\omega$, so the operator $A^*A + C^*C$ is of Schrödinger type:

$$(A^*A + C^*C)\,\omega = (-\partial_x^2 \omega + x^2 \omega - \omega) + \alpha^2 f'^2 \omega,$$

and the spectrum of $A^*A + C^*C$ can be studied via standard semi-classical techniques. For instance, if $f'(x)^2 = x^2/(1+x^2)^k$, $k \in \mathbb{N}$, then the real part of the spectrum of $A^*A + C^*C$ is bounded below like $O(|\alpha|^{2\nu})$, with $\nu = \min(1, 2/k)$. Then a careful examination of the proof of Theorem 18 yields a lower bound like $O(|\alpha|^\nu)$ on the real part of the spectrum of $A^*A + B$.

Tuning parameters in the appropriate way leads to optimal results in this problem; see [24] for more information. In this reference the authors also compare the results obtained by hypocoercivity, to the results obtained by a more precise (but more tricky) spectral analysis.

Part II

The auxiliary operator method

In this part I shall present an abstract hypocoercivity theorem applying to a linear operator L whose symmetric part is nonnegative, but which does not necessarily take the form $A^*A + B$. Still it will be useful to decompose L into its symmetric part S and its antisymmetric part B. Of course, we could always define A to be the square root of S, but this might be an extremely complicated operator, and the assumptions of the $A^*A + B$ Theorems might in practice be impossible to check. Important applications arise when the operator S contains an integral part, as in the linearized Boltzmann equation.

A classical general trick in spectral theory, when one studies the properties of a given linear operator L, consists in introducing an auxiliary operator which has good commutation properties with L. Here the idea will be similar, with just an important twist: We shall look for an auxiliary operator A which "almost commutes" with S and "does not at all" commute with B, in the sense that the effect of the commutator $[A, B]$ will be strong enough to enforce the coercivity of $S + [A, B]^*[A, B]$.

With this idea in mind, I had been looking for a hypocoercivity theorem generalizing, say, Theorem 18, but stumbled on the problem of practical verification of my assumptions. In the meantime, Clément Mouhot and Lukas Neumann found a theorem which, while in the same spirit of Theorem 18, has some important structural differences. The Mouhot–Neumann theorem is quite simple and turns out to be applicable to many important cases, as investigated in [**43**]; so in the sequel I shall only present their approach, with just slight variations and a more abstract treatment. Then I shall discuss the weak points of this method, and explain why another theory still needs to be developed, probably with slightly more sophisticated tools. At the time of writing, Frédéric Hérau has made partial progress in this direction.

10. Assumptions

In the sequel, $\mathcal{H}$ is a separable Hilbert space on $\mathbb{R}$ or $\mathbb{C}$, S is a nonnegative symmetric, possibly unbounded operator $\mathcal{H} \to \mathcal{H}$ and B is an antisymmetric, possibly unbounded operator $\mathcal{H} \to \mathcal{H}$. Then $A = (A_1, \ldots, A_m)$ is an array of unbounded operators $\mathcal{H} \to \mathcal{H}$. All of these operators are defined on a common dense domain. I shall actually ignore all regularity issues and be content with formal calculations, to be considered as a priori estimates.

The same conventions as in Section 1 will apply. Some of the assumptions below will involve $\sqrt{S}^{-1} U$ for various operators U; of course, this is not rigorous since $\sqrt{S}$ is in general not invertible. To make sense of these assumptions, one can either consider them as a priori estimates for a regularized problem in which S is replaced by an invertible approximation (something like $S + \varepsilon I$, and one tries to get estimates which are independent of ε); or supply them with the condition that $\sqrt{S}$ is invertible on the range of U (a trivial case of application is when $U = 0$).

The object of interest is the semigroup generated by the operator
$$L = S + B.$$
The next hypocoercivity theorem for L will make crucial use of the commutator of A and B. I shall write
$$[A, B] = ZC + R,$$
where Z is bounded from above and below, and R is some "remainder".

Now come a bunch of commutator conditions which will be used in Section 11. Later in Section 12 I shall make some simplifying assumptions which will drastically reduce the number of these conditions; but for the moment I shall keep the discussion at a general level.

(A1) $\begin{cases} \text{either} & [C, S] \preccurlyeq \sqrt{S} \\ \text{or} & \sqrt{S}^{-1}[C, S] \preccurlyeq \sqrt{S} \end{cases}$

(A2) $\begin{cases} \text{either} & (A \preccurlyeq \sqrt{S}A, C, \sqrt{S}C, \sqrt{S}) \\ & \text{and} \quad ([C, L] \preccurlyeq \sqrt{S}, \sqrt{S}C) \\ \text{or} & \sqrt{S}^{-1}[C, L] \preccurlyeq \sqrt{S}, \sqrt{S}C \end{cases}$

(A3) $\sqrt{S}[A^*, C] \preccurlyeq \sqrt{S}, \sqrt{S}C, C, \sqrt{S}A$

(A4) $(\sqrt{S}A^* \preccurlyeq \sqrt{S}A, C, \sqrt{S}C, \sqrt{S})$ and $(\sqrt{S}C^* \preccurlyeq \sqrt{S}C, \sqrt{S})$

(A5) $\begin{cases} \text{either} & (A^* \preccurlyeq \sqrt{S}A, C, \sqrt{S}C, \sqrt{S}) \\ & \text{and} \quad ([C^*, S] \preccurlyeq \sqrt{S}, \sqrt{S}C) \\ \text{or} & (\sqrt{S}A^* \preccurlyeq \sqrt{S}A, C, \sqrt{S}C, S) \\ & \text{and} \quad (\sqrt{S}^{-1}[C^*, S] \preccurlyeq \sqrt{S}, \sqrt{S}C) \\ \text{or} & [C^*, S] = 0 \end{cases}$

(A6) $R \preccurlyeq \sqrt{S}, \sqrt{S}C.$

(A7) There exist constants $\kappa, \bar{c} > 0$ such that for all $h \in \mathcal{H}$,
$$\langle Ah, ASh \rangle \geq \kappa \langle SAh, Ah \rangle - \bar{c}\Big(\langle Sh, h \rangle + \langle SCh, Ch \rangle + \|Ch\|^2\Big).$$

Here is a simple, but sometimes too restrictive, sufficient condition for **(A7)** to hold (the proof is left to the reader):

$$\textbf{(A7')} \begin{cases} \text{either} & (A \preccurlyeq \sqrt{S}A,\ C,\ \sqrt{S}C,\ \sqrt{S}) \\ & \text{and} \quad ([S,A] \preccurlyeq C, \sqrt{S}C, \sqrt{S}) \\ \text{or} & \sqrt{S}^{-1}[S,A] \preccurlyeq \sqrt{S},\ \sqrt{S}C,\ C \end{cases}$$

REMARK 44. Some of the assumptions **(A1)**–**(A7)** can be replaced by other assumptions involving the commutator $[A, S]$. I did not mention these alternative assumptions since they are in general more tricky to check that the ones which I chose. In case of need, the reader can easily find them by adapting the proof of the main theorem below.

11. Main theorem

THEOREM 45 (hypocoercivity for $L = S + B$). *With the same notation as in Section 10, assume that* **(A1)**–**(A7)** *are satisfied. Further assume that*

$$(11.1) \begin{cases} (i) & \exists \kappa, c > 0;\quad \forall h \in (\operatorname{Ker} L)^\perp, \\ & \langle ASh, Ah \rangle \geq \kappa \|Ah\|^2 \ -\ c\big(\langle Sh, h \rangle + \langle SCh, Ch \rangle + \|Ch\|^2\big); \\ (ii) & S + A^*SA + C^*SC + A^*A + C^*C \ \text{ is coercive on } (\operatorname{Ker} L)^\perp. \end{cases}$$

Then there are constants $c, \lambda > 0$, *depending only on the constants appearing implicitly in* **(A1)**–**(A7)** *and* (11.1), *such that*

$$\left\| e^{-tL} \right\|_{\widetilde{\mathcal{H}} \to \widetilde{\mathcal{H}}} \leq c\, e^{-\lambda t},$$

where $\widetilde{\mathcal{H}} \subset (\operatorname{Ker} L)^\perp$ *is defined by the Hilbert norm*

$$\|h\|_{\widetilde{\mathcal{H}}}^2 = \|h\|^2 + \|Ah\|^2 + \|Ch\|^2.$$

REMARK 46. Although Condition (11.1)(i) formally resembles Assumption **(A7)**, I have preferred to state it together with (11.1)(ii) because its practical verification often depends on a control of $\|h\|^2$ by $\langle Sh, h \rangle + \|Ch\|^2$.

PROOF OF THEOREM 45. The proof is quite similar in spirit to the proof of Theorem 18, so I shall be sketchy and only point out the main steps in the calculations.

First note that $(\operatorname{Ker} L)^\perp$ is stable under the evolution by e^{-tL}. Indeed, if $k \in \operatorname{Ker} L$, then $(d/dt)\langle e^{-tL}h, k \rangle = \langle Le^{-tL}h, k \rangle = \langle e^{-tL}h, L^*k \rangle$, so it is sufficient to show that $L^*k = 0$. But $Lk = 0$ implies $\langle Sk, k \rangle = \langle Lk, k \rangle = 0$, so $k \in \operatorname{Ker} S$ (here the nonnegativity of S is essential), so $k \in \operatorname{Ker} B$ also, and $L^*k = (S - B)k = 0$.

Next let

$$\mathcal{F}(h) = \|h\|^2 + a\|Ch\|^2 + 2b\, \Re \langle Ch, Ah \rangle + c\|Ah\|^2,$$

where $\Re$ stands for real part, and a, b, c will be chosen later in such a way that $1 \gg a \gg b \gg c > 0$, $a \ll \sqrt{b}$, $b \ll \sqrt{ac}$. In particular, $\mathcal{F}(h)$ will be bounded from above and below by constant multiples of $\|h\|_{\widetilde{\mathcal{H}}}^2$; so to prove the theorem it is sufficient to establish the estimate $(-d/dt)\mathcal{F}(e^{-tL}h) \geq \operatorname{const.}\mathcal{F}(e^{-tL}h)$. Without loss of generality, we can do it for $t = 0$ only. In the sequel, I shall also pretend that $\mathcal{H}$ is a real Hilbert space, so I shall not write real parts.

11. MAIN THEOREM

By direct computation,

(11.2) $$-\frac{d}{2\,dt}\bigg|_{t=0} \mathcal{F}(e^{-tL}h) = \langle Sh, h\rangle$$
$$+ a\langle CSh, Ch\rangle + a\langle CBh, Ch\rangle$$
$$+ b\langle CLh, Ah\rangle + b\langle Ch, ALh\rangle$$
$$+ c\langle ASh, Ah\rangle + c\langle ABh, Ah\rangle.$$

Now we shall estimate (11.2) line after line.

(1) The first line of (11.2) is kept unchanged.

(2) The second line of (11.2) is rewritten as follows:

(11.3) $$a\langle CSh, Ch\rangle = a\langle SCh, Ch\rangle + a\langle [C,S]h, Ch\rangle.$$

Then the second term in the right-hand side of (11.3) is estimated from below, either by $-a\|[C,S]h\|\,\|Ch\|$, or by $-a\|\sqrt{S}^{-1}[C,S]h\|\,\|\sqrt{S}Ch\|$; by Assumption (**A1**), these expressions can in turn be estimated from below by a constant multiple of

$$-a\Big(\|\sqrt{S}h\|\,\|Ch\| + \|\sqrt{S}h\|\,\|\sqrt{S}Ch\|\Big).$$

(Here I used the identity $\langle Su, u\rangle = \|\sqrt{S}u\|^2$.)

(3) The treatment of the third line of (11.2) is crucial; this is where the added coercivity from the commutator $[A, B]$ will show up. To handle the first term in this line, we write

$$\langle CLh, Ah\rangle = \langle LCh, Ah\rangle + \langle [C,L]h, Ah\rangle$$
$$= \langle SCh, Ah\rangle + \langle BCh, Ah\rangle + \langle [C,L]h, Ah\rangle$$
$$= \langle SCh, Ah\rangle - \langle Ch, BAh\rangle + \langle [C,L]h, Ah\rangle.$$

When we add this to the second term of the third line, $\langle Ch, ALh\rangle = \langle Ch, ABh\rangle + \langle Ch, ASh\rangle$, we obtain

$$\langle Ch, (AB-BA)h\rangle + \langle SCh, Ah\rangle + \langle [C,L]h, Ah\rangle + \langle Ch, ASh\rangle$$
$$= \langle Ch, (ZC+R)h\rangle + \langle SCh, Ah\rangle + \langle [C,L]h, Ah\rangle + \langle Ch, ASh\rangle$$
$$\geq \kappa\|Ch\|^2 + \langle Ch, Rh\rangle + \langle SCh, Ah\rangle + \langle [C,L]h, Ah\rangle + \langle Ch, ASh\rangle.$$

So there are four "error" terms to estimate from below:

(11.4) $\qquad \langle Ch, Rh\rangle, \qquad \langle SCh, Ah\rangle, \qquad \langle [C,L]h, Ah\rangle, \qquad \langle Ch, ASh\rangle.$

- To estimate the first term in (11.4), just write

$$\langle Ch, Rh\rangle \geq -\|Ch\|\,\|Rh\|$$

and apply Assumption (**A6**); there follows a lower bound by a constant multiple of

$$-b\|Ch\|\,\big(\|\sqrt{S}h\| + \|\sqrt{S}Ch\|\big).$$

- To estimate the second term in (11.4), use the Cauchy–Schwarz inequality:

$$\langle SCh, Ah\rangle \geq -\|\sqrt{S}Ch\|\,\|\sqrt{S}Ah\|.$$

- To estimate the third term in (11.4), write either

$$\langle [C,L]h, Ah\rangle \geq -\|[C,L]h\|\,\|Ah\|$$

or

$$\langle [C,L]h, Ah\rangle \geq -\|\sqrt{S}^{-1}[C,L]h\|\,\|\sqrt{S}Ah\|$$

and apply Assumption **(A2)**. It results a lower bound by a constant multiple of

$$-b(\|\sqrt{S}Ah\| + \|Ch\| + \|\sqrt{S}Ch\| + \sqrt{S}h\|)(\|\sqrt{S}h\| + \|\sqrt{S}Ch\|)$$
$$- b\left(\|\sqrt{S}h\| + \|\sqrt{S}Ch\|\right)\|\sqrt{S}Ah\|).$$

- The fourth term in (11.4) is a bit more tricky:

$$\begin{aligned}\langle Ch, ASh\rangle &= \langle A^*Ch, Sh\rangle \\ &= \langle [A^*, C]h, Sh\rangle + \langle CA^*h, Sh\rangle \\ &= \langle [A^*, C]h, Sh\rangle + \langle A^*h, C^*Sh\rangle \\ &= \langle [A^*, C]h, Sh\rangle + \langle A^*h, SC^*h\rangle + \langle A^*h, [C^*, S]h\rangle.\end{aligned}$$

This gives rise to three more terms to estimate:

(11.5) $\qquad \langle [A^*, C]h, Sh\rangle, \quad \langle A^*h, SC^*h\rangle, \quad \langle A^*h, [C^*, S]h\rangle.$

- To handle the first term in (11.5), write

$$\langle [A^*, C]h, Sh\rangle = \langle \sqrt{S}[A^*, C]h, \sqrt{S}h\rangle \geq -\|\sqrt{S}[A^*, C]h\|\,\|\sqrt{S}h\|;$$

then apply Assumption **(A3)** to bound $\|\sqrt{S}[A^*, C]h\|$. The result is a lower bound by a constant multiple of

$$-b\|\sqrt{S}h\|(\|\sqrt{S}h\| + \|\sqrt{S}Ch\| + \|Ch\| + \|\sqrt{S}Ah\|).$$

- To bound the second term in (11.5), write

$$\langle A^*h, SC^*h\rangle = \langle \sqrt{S}A^*h, \sqrt{S}C^*h\rangle \geq -\|\sqrt{S}A^*h\|\,\|\sqrt{S}C^*h\|;$$

then apply Assumption **(A4)** to bound these two norms separately. The result is a lower bound by a constant multiple of

$$-b(\|\sqrt{S}Ah\| + \|Ch\| + \|\sqrt{S}Ch\| + \|\sqrt{S}h\|)(\|\sqrt{S}Ch\| + \|\sqrt{S}h\|).$$

- To bound the last term in (11.5), one possibility is to write

$$\langle A^*h, [C^*, S]h\rangle \geq -\|A^*h\|\,\|[C^*, S]h\|;$$

another possibility is

$$\langle A^*h, [C^*, S]h\rangle = \langle \sqrt{S}A^*h, \sqrt{S}^{-1}[C^*, S]h\rangle \geq -\|\sqrt{S}A^*h\|\,\|\sqrt{S}^{-1}[C^*, S]h\|.$$

Then one can apply Assumption **(A5)** to control these terms. In the end, this gives a lower bound by a constant multiple of

$$-b(\|\sqrt{S}Ah\| + \|Ch\| + \|\sqrt{S}Ch\| + \|\sqrt{S}h\|)(\|\sqrt{S}h\| + \|\sqrt{S}Ch\|).$$

(4) Finally, the fourth line of (11.2) is handled as follows:

(11.6) $\quad \langle Ah, ASh\rangle + \langle Ah, ABh\rangle = \alpha\langle Ah, ASh\rangle + \beta\langle Ah, ASh\rangle$
$$+ \langle Ah, BAh\rangle + \langle Ah, [A, B]h\rangle,$$

where $\alpha, \beta \geq 0$ and $\alpha + \beta = 1$. The first term $\alpha\langle Ah, ASh\rangle$ is estimated by means of Assumption **(A7)**; the second term $\beta\langle Ah, ASh\rangle$ by means of Assumption (11.1)(i); altogether, these first two terms can be bounded below by a constant multiple of

$$c(\|\sqrt{S}Ah\| + \|Ah\|^2) - c(\|\sqrt{S}h\| + \|\sqrt{S}Ch\| + \|Ch\|^2).$$

Then the third term $\langle Ah, BAh\rangle$ in (11.6) vanishes; and the last term $\langle Ah, [A, B]h\rangle$ is bounded below by $-\|Ah\|\,\|ZCh\| - \|Ah\|\,\|Rh\|$, which in view of Assumption **(A6)** can be bounded below by a constant multiple of

$$-c\|Ah\|\,\|Ch\| - c\|Ah\|\,(\|\sqrt{S}h\| + \|\sqrt{S}Ch\|).$$

Gathering up all these lower bounds, we see that

$$-\frac{d}{2\,dt}\mathcal{F} \geq \text{const.}\,\langle X, mX\rangle,$$

where

$$X = \left(\|\sqrt{S}h\|,\, \|\sqrt{S}Ch\|,\, \|Ch\|,\, \|\sqrt{S}Ah\|,\, \|Ah\|\right),$$

m is the 5×5 matrix

$$m = \begin{bmatrix} 1 - Mb - Mc & -Ma - Mb & -Ma - Mb & -Mb & -Mc \\ 0 & a - Mb - Mc & -Mb & -Mb & -Mc \\ 0 & 0 & b - Mc & 0 & -Mc \\ 0 & 0 & 0 & c & 0 \\ 0 & 0 & 0 & 0 & c \end{bmatrix},$$

and M is a large number depending on the bounds appearing in the assumptions of the theorem.

Then by reasoning as in Section 4 and using Lemma A.22, we can find coefficients $a, b, c > 0$ and a constant $\kappa > 0$ such that

$$(11.7) \quad -\frac{d}{2\,dt}\bigg|_{t=0} \mathcal{F}(e^{-tL}h)$$
$$\geq \kappa\Big(\langle Sh, h\rangle + \langle SCh, Ch\rangle + \|Ch\|^2 + \langle SAh, Ah\rangle + \|Ah\|^2\Big).$$

By Assumption (11.1)(ii), this implies the existence of $\kappa', \kappa'' > 0$ such that

$$-\frac{d}{2\,dt}\bigg|_{t=0} \mathcal{F}(e^{-tL}h) \geq \kappa'\Big(\|h\|^2 + \langle Sh, h\rangle + \langle SCh, Ch\rangle + \|Ch\|^2$$
$$+ \langle SAh, Ah\rangle + \|Ah\|^2\Big)$$
$$\geq \kappa''\mathcal{F}(h).$$

This concludes the proof. $\square$

12. Simplified theorem and applications

In this section I shall consider a simplified version of Theorem 45.

COROLLARY 47. *Let $A = (A_1, \ldots, A_m), B, S$ be linear operators on a Hilbert space $\mathcal{H}$, and let $C = [A, B]$. Assume that*

$$A^* = -A, \quad B^* = -B, \quad C^* = -C, \quad S^* = S \geq 0;$$
$$[C, A] = 0, \quad [C, B] = 0, \quad [C, S] = 0.$$

Further assume that there exists $\kappa, c > 0$ such that for all $h \in (\operatorname{Ker} A \cap \operatorname{Ker} B)^\perp$,

$$(12.1) \quad \langle Ah, ASh\rangle \geq \kappa\big(\langle SAh, Ah\rangle + \|Ah\|^2\big)$$
$$- c\big(\langle Sh, h\rangle + \langle SCh, Ch\rangle + \|Ch\|^2\big);$$

and that

(12.2) $\qquad S + C^*C \quad$ *is coercive on* $(\operatorname{Ker} A \cap \operatorname{Ker} B)^\perp.$

Then there exists $\lambda > 0$ *such that*
$$\|e^{-t(S+B)}\|_{\widetilde{\mathcal{H}} \to \widetilde{\mathcal{H}}} = O(e^{-\lambda t}),$$
where $\widetilde{\mathcal{H}} \subset (\operatorname{Ker} L)^\perp$ *is defined by the Hilbert norm*
$$\|h\|_{\widetilde{\mathcal{H}}}^2 = \|h\|^2 + \|Ah\|^2 + \|Ch\|^2.$$

PROOF OF COROLLARY 47. The assumptions of the theorem trivially imply assumptions **(A1)**–**(A6)** from Section 10. Assumption 12.1 is equivalent to the conjunction of **(A7)** and (11.1)(i). Finally, (12.2) is obviously stronger than (11.1)(ii). □

Now let us make the link with the Mouhot–Neumann hypocoercivity theorem [**43**, Theorem 1.1]. Although the set of assumptions in that reference is not exactly the same as in the current section, we shall see that under a small additional hypothesis, the assumptions in [**43**] imply the present ones.

In [**43**], the Hilbert space $\mathcal{H}$ is $L^2(\mathbb{T}_x^n \times \mathbb{R}_v^n)$, and $A = \nabla_v$, $B = v \cdot \nabla_x$, $C = \nabla_x$; and the operator S only acts on the velocity variable v, so we have indeed $A^* = -A$, $B^* = -B$, $C^* = -C$, and C commutes with A, B and S. The kernel of L is similar to the kernel of S (up to identifying $v \to h(v)$ with $(x,v) \to h(v)$), and contains constant functions. Since $C^*C = -\Delta_x$ has a spectral gap, Condition (12.2) is equivalent to the fact that S has a spectral gap in $L^2(\mathbb{R}_v^n)$, which is Assumption H.3 in [**43**]. So it only remains to check (12.1), which will be true as soon as

(12.3) $\qquad \langle \nabla_v h, \nabla_v S h \rangle \geq \kappa\big(\langle S \nabla_v h, \nabla_v h\rangle + \|\nabla_v h\|^2\big) - c\|h\|^2.$

It is assumed in [**43**] that S, viewed as an operator on $L^2(\mathbb{R}_v^n)$, can be decomposed into the difference of two self-adjoint operators: $S = \Lambda - K$, where Λ is positive definite and

(12.4) $\qquad \langle \nabla_v h, \nabla_v \Lambda h \rangle \geq \kappa \langle \nabla_v h, \Lambda \nabla_v h \rangle - c\|h\|^2;$

(12.5) $\qquad \forall \delta > 0, \quad \exists c(\delta) > 0; \qquad \langle \nabla_v h, \nabla_v K h \rangle \leq \delta \|\nabla_v h\|^2 + c(\delta)\|h\|^2.$

Let us further assume that K *is compact relatively to* Λ, in the sense that
$$\forall \varepsilon > 0, \quad \exists c(\varepsilon) > 0; \qquad K \leq \varepsilon \Lambda + c(\varepsilon) I,$$
or equivalently (since $\Lambda = S + K$)

(12.6) $\qquad \forall \varepsilon > 0, \quad \exists c(\varepsilon) > 0; \qquad K \leq \varepsilon S + c(\varepsilon) I.$

By using (12.4), (12.5) and (12.6), and denoting by c and κ various positive constants, one easily obtains

$$\begin{aligned}\langle \nabla_v h, \nabla_v S h\rangle &= \langle \nabla_v h, \nabla_v \Lambda h\rangle - \langle \nabla_v h, \nabla_v K h\rangle \\ &\geq \kappa \langle \nabla_v h, \Lambda \nabla_v h\rangle - c\|h\|^2 - \langle \nabla_v h, \nabla_v K h\rangle \\ &\geq \kappa\big(\langle \nabla_v h, \Lambda \nabla_v h\rangle + \|\nabla_v h\|^2\big) - c\|h\|^2 - (\kappa/2)\|\nabla_v h\|^2 - c\|h\|^2 \\ &\geq \kappa\big(\langle \nabla_v h, \Lambda \nabla_v h\rangle + \|\nabla_v h\|^2\big) - c\|h\|^2 \\ &\geq \kappa\big(\langle \nabla_v h, S \nabla_v h\rangle + \|\nabla_v h\|^2\big) - c\big(\|h\|^2 + \langle \nabla_v h, K \nabla_v h\rangle\big) \\ &\geq \kappa\big(\langle \nabla_v h, S \nabla_v h\rangle + \|\nabla_v h\|^2\big) - c\|h\|^2.\end{aligned}$$

This establishes (12.3).

Assumption (12.6) is not made in [**43**], but it is satisfied in all the examples discussed therein: linear relaxation, semi-classical relaxation, linear Fokker–Planck equation, Boltzmann and Landau equations for hard potentials. So all these examples can be treated by means of Theorem 47. I refer to [**43**] for more explanations and results about all these models. Mouhot and Neumann also use these hypocoercivity results to construct smooth solutions for the corresponding *nonlinear* models close to equilibrium, thereby simplifying parts of the theory developed by Guo, see e.g. [**28**].

13. Discussion and open problems

Although it already applies to a number of interesting models, Theorem 45 suffers from several shortcomings. Consider for instance the case when S is a bounded operator (as in, say, the linearized Boltzmann equation for Maxwellian cross-section), and there is a force term $-\nabla V(x) \cdot \nabla_v$ in the left-hand side of the equation. Then the higher derivative term in $[C, L]$ is $-\nabla^2 V(x) \cdot \nabla_v$, which certainly *cannot* be bounded in terms of S and C; so Assumption (**A2**) does not hold. It is likely that Theorem 45 rarely applies in practice when $[C, B] = [[A, B], B] \neq 0$.

Other problems are due to Assumption (**A4**). This assumption will not hold for, say, $C = \nabla_x$ in a bounded domain $\Omega \subset \mathbb{R}^n$; indeed, in a slightly informal writing, $C^* = -C + \sigma \cdot dS$, where σ is the outer unit normal vector on $\partial\Omega$ and dS is the surface measure on $\partial\Omega$. So the computation used in the proof of Theorem 45 does not seem to give any result in such a situation.

A last indication that Theorem 45 is not fully satisfactory is that it does not seem to contain Theorem 18 as a particular case, although we would like to have a unified treatment of the general case $L = S + B$ and the particular case $L = A^*A + B$. In fact, as the reader may have noticed, the choices of coefficients in the auxiliary functionals appearing respectively in the proof of Theorem 18 and in the proof of Theorem 45 go in the opposite way!! Indeed, in the first case it was $\|h\|^2 + a\|Ah\|^2 + 2b\langle Ah, Ch\rangle + c\|Ch\|^2$ with $a \gg b \gg c$, while in the second case it was $\|h\|^2 + a\|Ch\|^2 + 2b\langle Ch, Ah\rangle + c\|Ah\|^2$.

Some playing around with the functionals suggests that these problems can be solved only if the auxiliary operator A is "*comparable*" to $\sqrt{S}$, say in terms of order of differential operators. So if S is bounded, then also A should be bounded. This suggests to modify the Mouhot–Neumann strategy in the case when S is bounded, by choosing, instead of $A = \nabla_v$, something like $A = (I - \Delta_v + v \cdot \nabla_v)^{-1/2} \nabla_v$. (I wrote $\Delta_v - v \cdot \nabla_v$ rather than Δ_v, because in many cases the natural reference measure is the Gaussian measure in $\mathbb{R}^n_v$.) Then computations involve nonlocal operators and become more intricate. I shall leave the problem open for future research.

Part III

Fully nonlinear equations

In this part I shall consider possibly nonlinear equations, and I shall not use "exact" commutator identities. To get significant results under such weak structure assumptions, I shall assume that I deal with solutions that are *very smooth*, uniformly in time. Moreover, I shall only prove results of convergence like $O(t^{-\infty})$, that is, faster than any inverse power of t.

As in Remark 14, the assumption of uniform smoothness can be relaxed as long as one has good estimates of exponential decay of singularities, together with a stability result (solutions depart from each other no faster than exponentially fast). However, I shall not address this issue here.

At the level of generality considered here, the rate $O(t^{-\infty})$ cannot be so much improved, since some cases are included for which exponential convergence simply does not hold, even for the linearized equation. In many situations one can still hope for rates of convergence like $O(e^{-\lambda t^\gamma})$, as in the close-to-equilibrium theory of the Boltzmann equation with soft potentials [**29**]. If a linearized study suggests convergence like $O(e^{-\lambda t})$ or $O(e^{-\lambda t^\gamma})$ for a particular nonlinear model, then one can try to obtain this rate of convergence by putting together the present nonlinear analysis (which applies far from equilibrium) with a linearization procedure (close to equilibrium) and a subsequent linear study.

This part is strongly influenced by my collaborations with Laurent Desvillettes on the convergence to equilibrium for the linear Fokker-Planck equation [**14**] and the nonlinear Boltzmann equation [**16**]. The method introduced in these papers was based on the study of second-order time differentiation of certain functionals; since then it has been successfully applied to other models [**9, 22**]. Our scheme of proof had several advantages: It was very general, physically meaningful, and gave us the intuition for the strong time-oscillations between hydrodynamic and homogeneous behavior, that were later observed numerically with a high accuracy [**23**]. On the other hand, our method had two major drawbacks: First, the heavy amount of calculations entailed by the second-order differentiations (especially in the presence of several conservation laws); and secondly, the particularly tricky analysis of the resulting coupled systems of second-order differential inequalities.

The approach will I shall adopt in the sequel remedies these drawbacks: First, it only uses first-order differentiation; secondly, it confines many heavy computations into a black box that can be used blindly. The price to pay will be the loss of intuition in the proof.

The main result is a rather abstract theorem stated in Section 14 and proven in Section 15. Then I shall show how to use this abstract result on various examples: the compressible Navier–Stokes system (Section 16); the Vlasov–Fokker–Planck equation with small smooth coupling (Section 17); and the Boltzmann equation (Section 18).

In the case of the Vlasov–Fokker–Planck equation to be considered, the coupling is simple enough that all the smoothness bounds appearing in the assumptions of the main theorem can be proven in terms of just assumptions on the initial data. In the other cases, the results will be conditional (depend on the validity of uniform regularity estimates).

The hard core of the proof of the main result was conceived during the conference "Advances in Mathematical Physics" in the honor of Carlo Cercignani (Montecatini, September 2004). It is a pleasure to thank the organizers of that meeting (Luigi Galgani, Maria Lampis, Rossanna Marra, Giuseppe Toscani) for helping to create a fruitful and pleasant atmosphere of work. The main results were first announced two weeks later, in an incomplete and preliminary form, at the Conference "Mathematical Aspects of Fluid and Plasma Dynamics" (Kyoto, September 2004), beautifully organized by Kazuo Aoki. During the Summer of 2006, for the purpose of various lectures in Porto Ercole, Trieste and Xining, I rewrote and generalized the main theorem, and added new applications. Additional thanks are due to Kazuo for an important remark about the treatment of the Boltzmann equation with Maxwellian diffusive boundary condition.

14. Main abstract theorem

The assumptions in this section are expressed in a rather abstract formalism. "Concrete" examples will be provided later in Sections 16 to 18.

14.1. Assumptions and main result. The theorem below involves five kinds of objects:

- a family of normed spaces $(X^s, \|\cdot\|_s)_{s\geq 0}$; the index s can be thought of as a way to quantify the regularity (smoothness, decay, etc.);

- two "differential" operators B and $\mathcal{C}$, such that B is "conservative" and $\mathcal{C}$ is "dissipative";

- a "very smooth" solution $t \to f(t)$ of the equation

$$\partial_t f + Bf = \mathcal{C}f,$$

with values in a subset X of the intersection of all the spaces X^s;

- a Lyapunov functional $\mathcal{E}$, which is dissipated by the equation above, and admits a unique absolute minimizer f_∞;

- a finite sequence of "nested nonlinear projections" $(\Pi_j)_{1 \leq j \leq J}$; one can think that Π_j is the projection onto the space of minimizers of $\mathcal{E}$ under $J - j$ constraints, and in particular Π_J is the map which takes everybody to f_∞.

The goal is to prove the convergence of $f(t)$ to the stationary state f_∞, and to get estimates on the rate of convergence.

I shall make several assumptions about these various objects. Even though these assumptions may look a bit lengthy and complicated, I tend to believe that they are satisfied in many natural cases. The following notation will be used:

- If A is an operator, then the image of a function f by A will be denoted either by $A(f)$ or simply by Af.

- The Fréchet derivative of A, evaluated at a function f, will be denoted by $A'(f)$ or A'_f; so $A'_f \cdot g$ stands for the Fréchet derivative of A evaluated at f and applied to the "tangent vector" g.

- The notation $\|A'(f)\|_{X \to Y}$ stands for the norm of the linear operator $A'(f) : X \to Y$, that is the smallest constant C such that $\|A'(f) \cdot g\|_Y \leq C\|g\|_X$ for all $g \in X$.

- Similarly, the second (functional) derivative of A, evaluated at a function f, will be denoted by $A''(f)$ or A''_f; so $A''_f \cdot (g, h)$ stands for the Hessian of A evaluated at f and applied to the two "tangent vectors" g and h. The notation $\|A''(f)\|_{X \to Y}$ stands for the smallest constant C such that $\|A''(f) \cdot (g, h)\|_Y \leq C\|g\|_X \|h\|_X$ for all $g, h \in X$.

ASSUMPTION 1 (scale of functional spaces). *$(X^s, \|\cdot\|_s)_{s\geq 0}$ is a nonincreasing family of Banach spaces such that*

(i) X^0 is Hilbert; its norm $\|\cdot\|_0$ will be denoted by just $\|\cdot\|$;

(ii) The injection $X^{s'} \subset X^s$ is continuous for $s' \geq s$; that is, there exists $C = C(s, s')$ such that

(14.1) $$\|f\|_s \leq C\|f\|_{s'};$$

(iii) The family $(X^s)_{s \geq 0}$ is an interpolation family: For any $s_0, s_1 \geq 0$ and $\theta \in [0, 1]$ there is a constant $C = C(s_0, s_1, \theta)$ such that

(14.2) $\quad s = (1-\theta)s_0 + \theta s_1 \implies \forall f \in X^{s_0} \cap X^{s_1}, \quad \|f\|_s \leq C\|f\|_{s_0}^{1-\theta}\|f\|_{s_1}^{\theta}.$

One may think of s as an index quantifying the regularity of f, say the number of derivatives which are bounded in a certain norm. In the sequel, I shall sometimes refer informally to s as an index standing for a number of derivatives, even if it is not necessarily so in general.

ASSUMPTION 2 (workspaces). X and Y are two sets such that $X \subset Y \subset \cap_{s \geq 0} X^s$; moreover, Y is convex and bounded in all spaces X^s.

ASSUMPTION 3 (solution). $f \in C(\mathbb{R}_+; X^s) \cap C^1((0, +\infty); X^s)$ for all s; moreover $f(t) \in X$ for all t. (In particular f is bounded in all spaces X^s.)

In the sequel, the notation f_0 will be a shorthand for $f(0)$.

ASSUMPTION 4 (equation). f solves the equation

(14.3) $$\frac{\partial f}{\partial t} + Bf = Cf,$$

where

(i) B, C are well-defined on Y and valued in a bounded subset of X^s for all s;

(ii) For any s there is s' large enough such that B' is bounded $X^{s'} \to X^s$, uniformly on Y;

(iii) C is Lipschitz $X^s \to X^0$, uniformly on Y, for s large enough.

In short, B and C satisfy a "Lipschitz condition with possible loss of derivatives". If X^s is a Sobolev space of order s on a bounded domain, then any reasonable differential operator of finite order, with smooth coefficients, will satisfy these assumptions.

ASSUMPTION 5 (stationary state). f_∞ is an element of X, satisfying $Bf_\infty = Cf_\infty = 0$.

ASSUMPTION 6 (projections). $(\Pi_j)_{1 \leq j \leq J}$ are nonlinear operators defined on Y, with $\Pi_J(Y) = \{f_\infty\}$. (Π_J sends everybody to the stationary state.) Moreover, for any $j \in \{1, \ldots, J\}$,

(i) $\Pi_j(X) \subset Y$, $\mathcal{C} \circ \Pi_j = 0$;

(ii) $\Pi_j f_\infty = f_\infty$;

(iii) For any s there is s' large enough such that $(\Pi_j)'$ and $(\Pi_j)''$ are bounded $X^{s'} \to X^s$, uniformly on Y.

The last of these assumptions morally says that Π_j is C^2 with possible loss of derivatives.

ASSUMPTION 7 (Lyapunov functional). $\mathcal{E} : Y \to \mathbb{R}$ is C^1 on Y viewed as a subset of X^s for s large enough. For all f one has $\mathcal{E}(f) \geq \mathcal{E}(\Pi_1 f) \geq \mathcal{E}(f_\infty)$, and more precisely

(i) For any $\varepsilon \in (0,1)$ there is $K_\varepsilon > 0$ such that for all $f \in Y$,
$$\mathcal{E}(f) - \mathcal{E}(\Pi_1 f) \geq K_\varepsilon \|f - \Pi_1 f\|^{2+\varepsilon}; \tag{14.4}$$

(ii) For any $\varepsilon \in (0,1)$ there are $K_\varepsilon, C_\varepsilon > 0$ such that for all $f \in Y$,
$$K_\varepsilon \|\Pi_1 f - f_\infty\|^{2+\varepsilon} \leq \mathcal{E}(\Pi_1 f) - \mathcal{E}(f_\infty) \leq C_\varepsilon \|\Pi_1 f - f_\infty\|^{2-\varepsilon}. \tag{14.5}$$

Note that $\Pi_1 f$ and f are bounded uniformly, so these bounds become more and more stringent as ε decreases.

ASSUMPTION 8 (Key hypocoercivity assumptions).
(i) $\mathcal{C}$ alone is dissipative, and strictly so out of the range of Π_1: For any $\varepsilon > 0$ there is a constant $K_\varepsilon > 0$ such that for all $f \in X$,
$$-\mathcal{E}'(f) \cdot (\mathcal{C}f) \geq K_\varepsilon \left[\mathcal{E}(f) - \mathcal{E}(\Pi_1 f)\right]^{1+\varepsilon}; \tag{14.6}$$

(ii) $\mathcal{C} - B$ is dissipative just as well: For any $\varepsilon > 0$ there is $K_\varepsilon > 0$ such that for all $f \in X$,
$$\mathcal{D}(f) := -\mathcal{E}'(f) \cdot (\mathcal{C}f - Bf) \geq K_\varepsilon \left[\mathcal{E}(f) - \mathcal{E}(\Pi_1 f)\right]^{1+\varepsilon}; \tag{14.7}$$

(iii) For any $k \leq J-1$ and for any $\varepsilon > 0$ there is a constant $K_\varepsilon > 0$ such that for all $f \in X$,
$$\mathcal{D}(f) + \sum_{j \leq k}\left\|(\mathrm{Id} - \Pi_j)'_{\Pi_j f} \cdot (B\Pi_j f)\right\|^2 \geq K_\varepsilon \|(\Pi_k - \Pi_{k+1})f\|^{2+\varepsilon}. \tag{14.8}$$

REMARK 48 (Simplified assumptions). In many cases of application, B is conservative, in the very weak sense that $\mathcal{E}'(f) \cdot (Bf) = 0$; then Assumption 8(ii) trivially follows from Assumption 8(i). Also most of the time, Assumption 8(iii) will be replaced by the stronger property
$$\left\|(\mathrm{Id} - \Pi_k)'_{\Pi_k f} \cdot (B\Pi_j f)\right\|^2 \geq K_\varepsilon \|(\Pi_k - \Pi_{k+1})f\|^{2+\varepsilon}. \tag{14.9}$$

In the sequel, I shall however discuss an important case (Boltzmann equation with Maxwellian diffuse boundary condition) where none of these simplifications holds true.

REMARK 49 (Practical verification of the key conditions). Often the Π_j's are *nested projectors*, in the sense that $\Pi_{j+1}\Pi_j = \Pi_{j+1}$. Then (14.9) becomes
$$\left\|(\mathrm{Id} - \Pi_k)'_g \cdot (Bg)\right\|^2 \geq K_\varepsilon \|(\mathrm{Id} - \Pi_{k+1})g\|^{2+\varepsilon}, \qquad g \in \Pi_k(X).$$

So the recipe is as follows: (a) Take $g \in \Pi_k(X)$, let it evolve according to $\partial_t g + Bg = 0$; (b) compute $\partial_t (\Pi_k g)$ at $t = 0$; and (c) check that $\|Bg + \partial_t (\Pi_k g)\|$ controls $\|g - \Pi_{k+1} g\|^{1+\varepsilon}$ for any $\varepsilon > 0$.

REMARK 50 (Connection with earlier works). To make the connection with the method used in [**16**], note that if $g = \Pi_k g$ at $t = 0$, then, since $\|\cdot\|$ is Hilbertian,
$$\left\|(\partial_t)_{t=0}(g_t - \Pi_k g_t)\right\|^2 = \frac{d^2}{dt^2}\bigg|_{t=0} \|g_t - \Pi_k g_t\|^2.$$

In view of this remark, Assumption 14.8 can be understood as a very abstract reformulation of the property of "instability of hydrodynamic description" introduced in [**16**].

Now comes the main nonlinear result in this memoir:

THEOREM 51. *Let Assumptions 1 to 8 be satisfied. Then, for any $\beta > 0$ there is a constant C_β, only depending on the constants appearing in these assumptions, on β and on an upper bound on $\mathcal{E}(f_0) - \mathcal{E}(f_\infty)$, such that*

$$\forall t \geq 0, \qquad \mathcal{E}(f(t)) - \mathcal{E}(f_\infty) \leq C_\beta\, t^{-\beta}.$$

As a consequence, for all $s \geq 0$,

$$\|f(t) - f_\infty\|_s = O(t^{-\infty}).$$

14.2. Method of proof. To estimate the speed of approach to equilibrium, the first natural thing to do is to consider the rate of decay of the Lyapunov functional $\mathcal{E}$. From the assumptions of Theorem 51, if $\varepsilon > 0$ is small enough then

$$(14.10) \qquad \frac{d}{dt}[\mathcal{E}(f) - \mathcal{E}(f_\infty)] = -\mathcal{D}(f) \leq -K_\varepsilon \big[\mathcal{E}(f) - \mathcal{E}(\Pi_1 f)\big]^{1+\varepsilon}.$$

(I have omitted the explicit dependence of f on t.) But the differential inequality (14.10) cannot in general be closed, since $\mathcal{E}(f) - \mathcal{E}(\Pi_1 f)$ might be much smaller than $\mathcal{E}(f) - \mathcal{E}(f_\infty)$. It may even be the case that $f = \Pi_1 f$, yet $f \neq f_\infty$ (the dissipation vanishes). So this strategy seems to be doomed.

In [14, 16] we solved this difficulty by coupling the differential inequality (14.10) with some second-order differential inequalities involving other functionals. Here on the contrary, I shall modify the functional $\mathcal{E}$ by adding some "lower-order" terms. So the proofs in the present paper are based on the following auxiliary functional:

$$(14.11) \qquad \mathcal{L}(f) = \big[\mathcal{E}(f) - \mathcal{E}(f_\infty)\big] + \sum_{j=1}^{J-1} a_j \Big\langle (\mathrm{Id} - \Pi_j)f,\, (\mathrm{Id} - \Pi_j)'_f \cdot (Bf) \Big\rangle,$$

where $\langle \cdot, \cdot \rangle$ denotes the scalar product in X^0, and $a_j > 0$ ($1 \leq j \leq J-1$) are carefully chosen small numbers, depending on smoothness bounds on f, and also on upper and lower bounds on $\mathcal{E}(f) - \mathcal{E}(f_\infty)$.

The coefficients a_j will be chosen in such a way that $\mathcal{L}(f)$ is always comparable to $\mathcal{E}(f) - \mathcal{E}(f_\infty)$; still the time-derivatives of these two quantities will be very different, and it will be possible to close the differential inequalities defined in terms of $\mathcal{L}$.

When the value of $\mathcal{E}(f) - \mathcal{E}(f_\infty)$ has substantially decreased, then the expression of $\mathcal{L}$ should be re-evaluated (the coefficients a_j should be updated), so $\mathcal{L}$ in itself does not really define a Lyapunov functional. But it will act just the same: On any time-interval where $\mathcal{E}(f) - \mathcal{E}(f_\infty)$ is controlled from above and below, one can choose the coefficients a_j in such a way that $(d/dt)\mathcal{L}(f) \leq -K\mathcal{L}(f)^{1+\delta}$, for any fixed $\delta > 0$. This will be sufficient to control the rate of decay of $\mathcal{L}$ to 0, and as a consequence the rate of decay of $\mathcal{E}$ to its minimum value.

Complete proofs will be given in the next section. It is clear that they enjoy some flexibility and can be slightly modified or adapted in case of need.

15. Proof of the Main Theorem

Theorem 51 will be obtained as a consequence of the following more precise result:

THEOREM 52. *Let Assumptions 1 to 8 be satisfied, and let $E > 0$ be such that*

$$(15.1) \qquad \frac{E}{2} \leq \mathcal{E}(f) - \mathcal{E}(f_\infty) \leq E.$$

Let further

$$\mathcal{L}(f) = \big[\mathcal{E}(f) - \mathcal{E}(f_\infty)\big] + \sum_{j=1}^{J-1} a_j \Big\langle (\mathrm{Id} - \Pi_j)f,\, (\mathrm{Id} - \Pi_j)'_f \cdot (Bf) \Big\rangle, \qquad (15.2)$$

where $(a_j)_{1 \leq j \leq J-1}$ are positive numbers; let $a_0 = 1$. Then,

(i) For any $\varepsilon \in (0,1)$, there is a constant $K > 0$, depending only on ε and on the constants appearing in Assumptions 1 to 8 (but not on E) such that if $a_j \leq K E^\varepsilon$ for all j, then

$$\forall f \in X, \qquad \frac{E}{4} \leq \mathcal{L}(f) \leq \frac{5E}{4}.$$

(ii) There are absolute constants $\varepsilon_0, k > 0$, and there are constants $K, K' > 0$, depending only on ε, on an upper bound on $\mathcal{E}(f) - \mathcal{E}(f_\infty)$ and on the constants appearing in Assumptions 1 to 8 such that, if $0 < \varepsilon \leq \varepsilon_0$ and

$$a_{j+1} \leq a_j, \qquad \frac{a_{j+1}^2}{a_j} \leq K\, a_{J-1}^{1+\varepsilon}\, E^{k\varepsilon},$$

for all $j \in \{0, \ldots, J-2\}$, then

$$\forall f \in X, \qquad \mathcal{L}'(f) \cdot (\mathcal{C}f - Bf) \leq -a_{J-1} K' E^{1+\varepsilon}.$$

REMARK 53. Lemma A.23 in Appendix A.23 shows that Conditions (i) and (ii) can be fulfilled with $a_{J-1} \geq K_1 E^{\ell\varepsilon}$, where ℓ only depends on J and k.

REMARK 54. In concrete situations, the explicit form of $\mathcal{L}$ might be extremely complicated. In the case of the Boltzmann equation, to be considered later on, the formula for $\mathcal{L}$ requires eight lines of display.

Before starting the proof of Theorem 52, let me make some remarks to facilitate its reading. First of all, when uniform bounds in the X^s spaces are taken for granted, a bound from above by, say, $\|f - f_\infty\|_s^\alpha$ is better if the exponent α is *higher*; this is somewhat contrary to what one is used to when working on smoothness a priori estimates.

In all the sequel the exponents s, s' and the constants C, C', K, K', etc. may change from one formula to the other. These quantities can all be computed in terms of an upper bound on $\mathcal{E}(f_0) - \mathcal{E}(f_\infty)$, the exponents and constants appearing in Assumptions 1 to 8 (and for given ε, they only involve a finite number of these constants and exponents). As a general rule, the symbols C, C', etc. will stand for constants which should be taken *large* enough, while the symbols K, K', etc. will stand for positive constants which should be taken *small* enough.

Finally, I shall frequently use the following fact: If $\|g\|_{s'} \leq C_{s'}$ for all $s' \geq 0$, then for any s and any δ there exists a constant C, only depending on $C_{s'}$ for some s' large enough, such that

$$\|g\|_s \leq C \|g\|_{s'}^{1-\delta}. \qquad (15.3)$$

To see this, it suffices to use (14.2) with $s_0 = 0$, $s_1 = s/\delta$, $\theta = \delta$. In other words, it is always possible to replace the norm in some X^s by the norm in any other X^s, up to a arbitrarily small deterioration of the exponents.

PROOF OF THEOREM 52. To prove (i), it is sufficient to show that there exists C such that

$$\Big| \big\langle (\mathrm{Id} - \Pi_j)f,\, (\mathrm{Id} - \Pi_j)'_f \cdot (Bf) \big\rangle \Big| \leq C \left[\mathcal{E}(f) - \mathcal{E}(f_\infty)\right]^{1-\varepsilon} \qquad (15.4)$$

for all $f \in X$. Indeed, it will follow from (15.1) that
$$\left|\left\langle (\mathrm{Id} - \Pi_j)f, (\mathrm{Id} - \Pi_j)'_f \cdot (Bf) \right\rangle\right| \leq \frac{2^\varepsilon C}{E^\varepsilon} [\mathcal{E}(f) - \mathcal{E}(f_\infty)];$$
then if $a_j \leq KE^\varepsilon$, the definition of $\mathcal{L}$ (formula (15.2)) will imply
$$(1 - 2^\varepsilon JKC)[\mathcal{E}(f) - \mathcal{E}(f_\infty)] \leq \mathcal{L}(f) \leq (1 + 2^\varepsilon JKC)[\mathcal{E}(f) - \mathcal{E}(f_\infty)].$$
Then the conclusion will be obtained by choosing, say, $K = 1/(2^{2+\varepsilon} JC)$. (Here C is the same constant as in (15.4).)

To prove (15.4), I shall first apply the Cauchy–Schwarz inequality, and bound separately $\|(\mathrm{Id} - \Pi_j)f\|$ and $\|(\mathrm{Id} - \Pi_j)'_f \cdot (Bf)\|$.

Bound on $\|(\mathrm{Id} - \Pi_j)f\|$:

By Assumption 6(ii), $f - \Pi_j f = (f - f_\infty) - (\Pi_j f - \Pi_j f_\infty)$, so
$$\|f - \Pi_j f\| \leq \|f - f_\infty\| + \|\Pi_j f - \Pi_j f_\infty\|.$$
By Assumption 6(iii) and the convexity of Y, Π_j is Lipschitz $X^s \to X^0$ for some s large enough; so
$$\|\Pi_j f - \Pi_j f_\infty\| \leq C\|f - f_\infty\|_s.$$
Both f and f_∞ belong to Y, so by Assumption 3 they are bounded in $X^{s'}$ for all s', and we can apply the interpolation inequality (15.3):
$$\|f - f_\infty\|_s \leq C\|f - f_\infty\|^{1-\frac{\varepsilon}{2}}.$$
Then by Assumption 7(i)-(ii),
$$\|f - f_\infty\|^{1-\frac{\varepsilon}{2}} \leq C[\mathcal{E}(f) - \mathcal{E}(f_\infty)]^{\frac{1}{2}-\frac{\varepsilon}{2}}.$$
All in all,
$$(15.5) \qquad \|f - \Pi_j f\| \leq C[\mathcal{E}(f) - \mathcal{E}(f_\infty)]^{\frac{1}{2}-\frac{\varepsilon}{2}}.$$
Bound on $\|(\mathrm{Id} - \Pi_j)'_f \cdot (Bf)\|$:

By Assumption 6(iii), there are constants C and s such that
$$\|(\mathrm{Id} - \Pi_j)'_f \cdot (Bf)\| \leq C\|Bf\|_s.$$
By Assumption 4(i), Bf is bounded in all spaces $X^{s'}$, so by interpolation,
$$\|Bf\|_s \leq C\|Bf\|^{1-\frac{\varepsilon}{4}}.$$
It follows from Assumption 4(ii) and the convexity of Y that B is Lipschitz $X^s \to X^0$ on Y; in view of Assumption 5 ($Bf_\infty = 0$), this leads to
$$\|Bf\|^{1-\frac{\varepsilon}{4}} = \|Bf - Bf_\infty\|^{1-\frac{\varepsilon}{4}} \leq C\|f - f_\infty\|_s^{1-\frac{\varepsilon}{4}}.$$
The end of the estimate is just as before:
$$\|f - f_\infty\|_s^{1-\frac{\varepsilon}{4}} \leq C\|f - f_\infty\|^{1-\frac{\varepsilon}{2}} \leq C'[\mathcal{E}(f) - \mathcal{E}(f_\infty)]^{\frac{1}{2}-\frac{\varepsilon}{2}}.$$
All in all,
$$\|(\mathrm{Id} - \Pi_j)'_f \cdot (Bf)\| \leq C[\mathcal{E}(f) - \mathcal{E}(f_\infty)]^{\frac{1}{2}-\frac{\varepsilon}{2}}.$$
This combined with (15.5) establishes (15.4).

Now we turn to the proof of (ii), which is considerably more tricky. Let
$$(15.6) \qquad \widetilde{\mathcal{D}}(f) := -\mathcal{L}(f) \cdot (\mathcal{C}f - Bf).$$
The argument will be divided in three steps.

Step 1: The estimates in this step are mainly based on regularity assumptions.

By direct computation,

$$-\widetilde{\mathcal{D}}(f) = -\mathcal{D}(f) + \sum_{j=1}^{J-1} a_j \big\langle (\mathrm{Id} - \Pi_j)'_f \cdot (\mathcal{C}f - Bf),\, (\mathrm{Id} - \Pi_j)'_f \cdot (Bf) \big\rangle$$

$$+ \sum_{j=1}^{J-1} a_j \big\langle (\mathrm{Id} - \Pi_j)f,\, (\mathrm{Id} - \Pi_j)''_f \cdot (\mathcal{C}f - Bf, Bf) \big\rangle$$

$$+ \sum_{j=1}^{J-1} a_j \big\langle (\mathrm{Id} - \Pi_j)f,\, (\mathrm{Id} - \Pi_j)'_f \cdot (B'_f \cdot (\mathcal{C}f - Bf)) \big\rangle$$

$$= -\mathcal{D}(f) - \sum_{j=1}^{J-1} a_j \big\| (\mathrm{Id} - \Pi_j)'_f \cdot (Bf) \big\|^2$$

$$+ \sum_{j=1}^{J-1} a_j \big\langle (\mathrm{Id} - \Pi_j)'_f \cdot (\mathcal{C}f),\, (\mathrm{Id} - \Pi_j)'_f \cdot (Bf) \big\rangle$$

$$+ \sum_{j=1}^{J-1} a_j \big\langle (\mathrm{Id} - \Pi_j)f,\, (\mathrm{Id} - \Pi_j)''_f \cdot (\mathcal{C}f - Bf, Bf) \big\rangle$$

$$+ \sum_{j=1}^{J-1} a_j \big\langle (\mathrm{Id} - \Pi_j)f,\, (\mathrm{Id} - \Pi_j)'_f \cdot (B'_f \cdot (\mathcal{C}f - Bf)) \big\rangle$$

Then by Cauchy-Schwarz inequality,

$$(15.7) \quad -\widetilde{\mathcal{D}}(f) \leq - \mathcal{D}(f) - \sum_{j=1}^{J-1} a_j \big\| (\mathrm{Id} - \Pi_j)'_f \cdot (Bf) \big\|^2$$

$$+ \sum_{j=1}^{J-1} a_j \big\| (\mathrm{Id} - \Pi_j)'_f \cdot (\mathcal{C}f) \big\| \big\| (\mathrm{Id} - \Pi_j)'_f \cdot (Bf) \big\|$$

$$+ \sum_{j=1}^{J-1} a_j \big\| (\mathrm{Id} - \Pi_j)f \big\| \big\| (\mathrm{Id} - \Pi_j)''_f \cdot (\mathcal{C}f - Bf, Bf) \big\|$$

$$+ \sum_{j=1}^{J-1} a_j \big\| (\mathrm{Id} - \Pi_j)f \big\| \, \big\| (\mathrm{Id} - \Pi_j)'_f \cdot (B'_f \cdot (\mathcal{C}f - Bf)) \big\|.$$

By the inequality $ab \leq (a^2 + b^2)/2$ with $a = \|(\mathrm{Id} - \Pi_j)'_f \cdot (Bf)\|$ and $b = \|(\mathrm{Id} - \Pi_j)'_f \cdot (\mathcal{C}f)\|$, we see that the second and third terms in the right-hand side of (15.7) can be bounded by

$$(15.8) \quad -\frac{1}{2} \sum_{j=1}^{J-1} a_j \big\| (\mathrm{Id} - \Pi_j)'_f \cdot (Bf) \big\|^2 + \frac{1}{2} \sum_{j=1}^{J-1} a_j \big\| (\mathrm{Id} - \Pi_j)'_f \cdot (\mathcal{C}f) \big\|^2.$$

Then we apply the Hilbertian inequality

$$-\|a\|^2 \leq -\frac{\|b\|^2}{2} + \|b - a\|^2$$

with $a = (\mathrm{Id} - \Pi_j)'_f \cdot (Bf)$ and $b = (\mathrm{Id} - \Pi_j)'_{\Pi_j f} \cdot (B\Pi_j f)$, to bound (15.8) by

$$-\frac{1}{4}\sum_{j=1}^{J-1} a_j \left\|(\mathrm{Id} - \Pi_j)'_{\Pi_j f} \cdot (B\Pi_j f)\right\|^2$$

$$+\frac{1}{2}\sum_{j=1}^{J-1} a_j \left\|(\mathrm{Id} - \Pi_j)'_{\Pi_j f} \cdot (B\Pi_j f) - (\mathrm{Id} - \Pi_j)'_f \cdot (Bf)\right\|^2$$

$$+\frac{1}{2}\sum_{j=1}^{J-1} a_j \left\|(\mathrm{Id} - \Pi_j)'_f \cdot (\mathcal{C}f)\right\|^2.$$

It follows, after plugging these bounds back in (15.7), that

(15.9) $\quad -\widetilde{\mathcal{D}}(f) \leq -\mathcal{D}(f) - \dfrac{1}{4}\displaystyle\sum_{j=1}^{J-1} a_j \left\|(\mathrm{Id} - \Pi_j)'_{\Pi_j f} \cdot (B\Pi_j f)\right\|^2 + \displaystyle\sum_{j=1}^{J-1} a_j \, (R)_j,$

where

(15.10) $\quad (R)_j := \dfrac{1}{2}\left\|(\mathrm{Id} - \Pi_j)'_{\Pi_j f} \cdot (B\Pi_j f) - (\mathrm{Id} - \Pi_j)'_f \cdot (Bf)\right\|^2$

$$+ \frac{1}{2}\left\|(\mathrm{Id} - \Pi_j)'_f \cdot (\mathcal{C}f)\right\|^2$$

$$+ \|(\mathrm{Id} - \Pi_j)f\|\Big(\left\|(\mathrm{Id} - \Pi_j)''_f \cdot (\mathcal{C}f - Bf, Bf)\right\|$$

$$+ \left\|(\mathrm{Id} - \Pi_j)'_f \cdot (B'_f \cdot (\mathcal{C}f - Bf))\right\|\Big).$$

Now I shall estimate the various terms in (15.10) one after the other.

First line of (15.10):

First,

(15.11) $\quad \left\|(\mathrm{Id} - \Pi_j)'_{\Pi_j f} \cdot (B\Pi_j f) - (\mathrm{Id} - \Pi_j)'_f \cdot (Bf)\right\| \leq$

$$\left\|(\mathrm{Id} - \Pi_j)'_{\Pi_j f} \cdot (Bf - B\Pi_j f)\right\| + \left\|[(\Pi_j)'_{\Pi_j f} - (\Pi_j)'_f] \cdot (Bf)\right\|.$$

By Assumption 6(iii),

$$\left\|(\mathrm{Id} - \Pi_j)'_{\Pi_j f} \cdot (Bf - B\Pi_j f)\right\| \leq C\|Bf - B\Pi_j f\|_s.$$

(Here I use the fact that $\Pi_j(X) \subset Y$.) Also, from Assumptions 3 and 4 (and again $\Pi_j(X) \subset Y$), Bf and $B\Pi_j f$ are bounded in all spaces $X^{s'}$, so by interpolation

$$\|Bf - B\Pi_j f\|_s \leq C\|Bf - B\Pi_j f\|^{1-\frac{\varepsilon}{2}}.$$

As a consequence of Assumption 4(ii) and the convexity of Y, B is Lipschitz continuous $X^s \to X^0$ on Y, so

$$\|Bf - B\Pi_j f\|^{1-\frac{\varepsilon}{2}} \leq C\|f - \Pi_j f\|_s^{1-\frac{\varepsilon}{2}} \leq C\|f - \Pi_j f\|^{1-\varepsilon},$$

where the last inequality is obtained again from interpolation. This provides a bound for the first term on the right-hand side of (15.11)

Next, as a consequence of Assumption 6(iii), $(\Pi_j)'$ is Lipschitz continuous on Y, in the sense that for all $f, g \in Y$,

$$\left\|[(\Pi_j)'_f - (\Pi_j)'_g] \cdot h\right\| \leq C\|f - g\|_s \|h\|_s.$$

Combining this with Assumption 4, we find

$$\left\|[(\Pi_j)'_{\Pi_j f} - (\Pi_j)'_f] \cdot (Bf)\right\| \leq C \|\Pi_j f - f\|_s \|Bf\|_s$$
$$\leq C' \|f - \Pi_j f\|_s \leq C'' \|f - \Pi_j f\|^{1-\varepsilon}.$$

This takes care of the second term on the right-hand side of (15.11). The conclusion is that the first line of (15.10) is bounded above by $O(\|f - \Pi_j f\|^{1-\varepsilon})$, for any $\varepsilon \in (0, 1)$.

Second line of (15.10):

First, by Assumption 6(iii),

$$\left\|(\mathrm{Id} - \Pi_j)'_f \cdot (\mathcal{C}f)\right\| \leq C \|\mathcal{C}f\|_s.$$

By Assumption 4(i), $\mathcal{C}f$ is bounded in all spaces $X^{s'}$, so by interpolation:

$$\|\mathcal{C}f\|_s \leq C \|\mathcal{C}f\|^{1-\frac{\varepsilon}{2}}.$$

By Assumption 6(i), $\mathcal{C}\Pi_j f = 0$; and by Assumption 4(iii), $\mathcal{C}$ is Lipschitz $X^s \to X^0$ on Y; so

$$\|\mathcal{C}f\|^{1-\frac{\varepsilon}{2}} = \|\mathcal{C}f - \mathcal{C}\Pi_j f\|^{1-\frac{\varepsilon}{2}} \leq C \|f - \Pi_j f\|_s^{1-\frac{\varepsilon}{2}} \leq C' \|f - \Pi_j f\|^{1-\varepsilon}.$$

The conclusion is that the second line of (15.10) can be bounded just as the first line, by $O(\|f - \Pi_j f\|^{1-\varepsilon})$, for any $\varepsilon \in (0, 1)$.

Third and fourth lines of (15.10):

By Assumption 6(iii),

$$\left\|(\mathrm{Id} - \Pi_j)''_f \cdot (\mathcal{C}f - Bf, Bf)\right\| \leq C \|\mathcal{C}f - Bf\|_s \|Bf\|_s \leq C' (\|Bf\|_s^2 + \|\mathcal{C}f\|_s^2).$$

The second term $\|\mathcal{C}f\|_s^2$ can be bounded by $O(\|f - \Pi_k f\|^{2-2\varepsilon})$, as we already saw; by taking $k = J$ we get a bound like $O(\|f - f_\infty\|^{2-\varepsilon})$. As for the first term $\|Bf\|_s^2$, we saw before that it is also bounded like $O(\|f - f_\infty\|^{2-\varepsilon})$. In the sequel I shall only keep the worse bound $O(\|f - f_\infty\|^{1-\varepsilon})$.

Next, by Assumption 6(iii) again,

$$\left\|(\mathrm{Id} - \Pi_j)'_f \cdot (B'_f \cdot (\mathcal{C}f - Bf))\right\| \leq C \|B'_f \cdot (\mathcal{C}f - Bf)\|_s.$$

From Assumption 4(ii),

$$\|B'_f \cdot (\mathcal{C}f - Bf)\|_s \leq C \|\mathcal{C}f - Bf\|_{s'} \leq C (\|\mathcal{C}f\|_{s'} + \|Bf\|_{s'}),$$

and as before this can be controlled by $O(\|f - f_\infty\|^{1-\varepsilon})$.

The conclusion is that the third and fourth lines of (15.10) can be bounded by $C\|f - \Pi_j f\| \|f - f_\infty\|^{1-\varepsilon}$, for any $\varepsilon \in (0, 1)$.

Gathering all these estimates and replacing ε by $\varepsilon/2$, we deduce that the expression in (15.10) can be bounded as follows:

$$(R)_j \leq C \left(\|f - \Pi_j f\|^{2-\frac{\varepsilon}{2}} + \|f - \Pi_j f\| \|f - f_\infty\|^{1-\frac{\varepsilon}{2}} \right).$$

As we already saw before,

$$\|f - \Pi_j f\|^{1-\frac{\varepsilon}{2}} \leq C \|f - f_\infty\|^{1-\varepsilon},$$

so actually

(15.12) $$(R)_j \leq C \|f - \Pi_j f\| \|f - f_\infty\|^{1-\varepsilon}.$$

The temporary conclusion is that

$$(15.13) \quad -\widetilde{\mathcal{D}}(f) \leq -\mathcal{D}(f) - \frac{1}{4}\sum_{j=1}^{J-1} a_j \left\|(\mathrm{Id}-\Pi_j)'_{\Pi_j f} \cdot (B\Pi_j f)\right\|^2$$

$$+ \left(\sum_{j=1}^{J-1} a_j \|f-\Pi_j f\|\right) \|f-f_\infty\|^{1-\varepsilon}.$$

Step 2: This step uses Assumption 8 crucially.

By triangle inequality,

$$\|f-\Pi_j f\| = \|\Pi_0 f - \Pi_j f\| \leq \sum_{0\leq k\leq j-1} \|\Pi_k f - \Pi_{k+1} f\|;$$

so

$$\sum_{1\leq j\leq J-1} a_j \|f-\Pi_j f\| \leq \sum_{1\leq j\leq J-1} a_j \left(\sum_{0\leq k\leq j-1} \|\Pi_k f - \Pi_{k+1} f\|\right)$$

$$= \sum_{0\leq k\leq J-2} \left(\sum_{k+1\leq j\leq J-1} a_j\right) \|\Pi_k f - \Pi_{k+1} f\|$$

$$\leq \sum_{0\leq k\leq J-2} (J a_{k+1}) \|\Pi_k f - \Pi_{k+1} f\|,$$

where the last inequality is due to the sequence $(a_j)_{0\leq j\leq J-1}$ being nonincreasing.

Renaming k as j, plugging this inequality back in (15.13), we arrive at

$$(15.14) \quad -\widetilde{\mathcal{D}}(f) \leq -\mathcal{D}(f) - \frac{1}{4}\sum_{0\leq j\leq J-2} a_j \left\|(\mathrm{Id}-\Pi_j)'_{\Pi_j f} \cdot (B\Pi_j f)\right\|^2$$

$$+ C \sum_{0\leq j\leq J-2} a_{j+1} \|\Pi_j f - \Pi_{j+1} f\| \|f-f_\infty\|^{1-\varepsilon}.$$

Since $a_k \leq 1$, we can write

$$(15.15) \quad \mathcal{D}(f) = \frac{\mathcal{D}(f)}{2} + \frac{1}{2(J-1)} \sum_{k=1}^{J-1} \mathcal{D}(f)$$

$$\geq \frac{\mathcal{D}(f)}{2} + \frac{1}{2(J-1)} \sum_{k=1}^{J-1} a_k \mathcal{D}(f).$$

On the other hand, $k \geq j \Longrightarrow a_j \geq a_k$, so

$$(15.16) \quad \sum_{j=1}^{J-1} a_j \left\|(\mathrm{Id}-\Pi_j)'_{\Pi_j f} \cdot (B\Pi_j f)\right\|^2$$

$$= \frac{1}{J-1} \sum_{j=1}^{J-1}\sum_{k=1}^{J-1} a_j \left\|(\mathrm{Id}-\Pi_j)'_{\Pi_j f} \cdot (B\Pi_j f)\right\|^2$$

$$\geq \frac{1}{J-1} \sum_{j=1}^{J-1}\sum_{k=1}^{J-1} a_k \left\|(\mathrm{Id}-\Pi_j)'_{\Pi_j f} \cdot (B\Pi_j f)\right\|^2.$$

From (15.14), (15.15) and (15.16),

$$-\widetilde{\mathcal{D}}(f) \leq -\frac{\mathcal{D}(f)}{2}$$
$$-\frac{1}{4(J-1)} \sum_{k=1}^{J-1} a_k \left(\mathcal{D}(f) + \sum_{j=1}^{k} \left\| (\mathrm{Id} - \Pi_j)'_{\Pi_j f} \cdot (B\Pi_j f) \right\|^2 \right)$$
$$+ C \sum_{j=0}^{J-2} a_{j+1} \left\| \Pi_j f - \Pi_{j+1} f \right\| \left\| f - f_\infty \right\|^{1-\varepsilon}.$$

At this point we can apply Assumption 8(ii)-(iii) and Assumption 7 and we get constants K, C such that

(15.17) $$-\widetilde{\mathcal{D}}(f) \leq - K \Big([\mathcal{E}(f) - \mathcal{E}(\Pi_1 f)]^{1+\frac{\varepsilon}{4}} + \left\| f - \Pi_1 f \right\|^{2+\varepsilon} \Big)$$
$$- K \sum_{k=1}^{J-1} a_k \left\| \Pi_k f - \Pi_{k+1} f \right\|^{2+\varepsilon}$$
$$+ C \sum_{j=0}^{J-2} a_{j+1} \left\| \Pi_j f - \Pi_{j+1} f \right\| \left\| f - f_\infty \right\|^{1-\varepsilon}.$$

By applying Young's inequality, in the form

$$a X Y^{1-\varepsilon} \leq b \frac{X^{2+\varepsilon}}{2+\varepsilon} + \left(\frac{a^{\frac{2+\varepsilon}{1+\varepsilon}}}{b^{\frac{1}{1+\varepsilon}}} \right) \frac{(Y^{1-\varepsilon})^{\left(\frac{2+\varepsilon}{1+\varepsilon}\right)}}{\left(\frac{2+\varepsilon}{1+\varepsilon}\right)},$$

with $a = a_{j+1}$, $X = \left\| \Pi_j f - \Pi_{j+1} f \right\|$, $Y = \left\| f - f_\infty \right\|$, $b = K a_j$ in the last line of (15.17), we get

(15.18) $$-\widetilde{\mathcal{D}}(f) \leq - K [\mathcal{E}(f) - \mathcal{E}(\Pi_1 f)]^{1+\frac{\varepsilon}{4}} - K \left\| f - \Pi_1 f \right\|^{2+\varepsilon}$$
$$- K \sum_{j=1}^{J-1} a_j \left\| \Pi_j f - \Pi_{j+1} f \right\|^{2+\varepsilon}$$
$$+ C \sum_{0 \leq j \leq J-2} \left(\frac{a_{j+1}^{2+\varepsilon}}{a_j} \right)^{\frac{1}{1+\varepsilon}} \left\| f - f_\infty \right\|^{\frac{(1-\varepsilon)(2+\varepsilon)}{1+\varepsilon}}.$$

Since $a_{j+1} \leq 1$, we can bound trivially $a_{j+1}^{2+\varepsilon}$ by a_{j+1}^2. Moreover, for $\varepsilon \leq \varepsilon_0$ small enough, we have $(1-\varepsilon)(2+\varepsilon)/(1+\varepsilon) \geq 2 - 4\varepsilon$ and, so

$$\left\| f - f_\infty \right\|^{\frac{(1-\varepsilon)(2+\varepsilon)}{1+\varepsilon}} \leq C \left\| f - f_\infty \right\|^{2-4\varepsilon}.$$

Taking into account once again the fact that $a_j \geq a_{J-1}$ for all j, (15.18) implies, with the convention $\Pi_0 f = f$, $a_0 = 1$,

(15.19) $$-\widetilde{\mathcal{D}}(f) \leq - K[\mathcal{E}(f) - \mathcal{E}(\Pi_1 f)]^{1+\frac{\varepsilon}{4}}$$
$$- K a_{J-1} \left(\sum_{0 \leq j \leq J-1} \left\| \Pi_j f - \Pi_{j+1} f \right\|^{2+\varepsilon} \right)$$
$$+ C \sup_{0 \leq j \leq J-2} \left(\frac{a_{j+1}^2}{a_j} \right)^{\frac{1}{1+\varepsilon}} \left\| f - f_\infty \right\|^{2-4\varepsilon}.$$

Next,
$$\|f - f_\infty\|^{2+\varepsilon} = \|\Pi_0 f - \Pi_J f\|^{2+\varepsilon} \leq C \sum_{0 \leq j \leq J-1} \|\Pi_j f - \Pi_{j+1} f\|^{2+\varepsilon},$$

so from (15.19) we deduce

(15.20) $\quad -\widetilde{\mathcal{D}}(f) \leq -K[\mathcal{E}(f) - \mathcal{E}(\Pi_1 f)]^{1+\frac{\varepsilon}{4}} - K\, a_{J-1} \|f - f_\infty\|^{2+\varepsilon}$
$$+ C \sup_{0 \leq j \leq J-2} \left(\frac{a_{j+1}^2}{a_j} \right)^{\frac{1}{1+\varepsilon}} \|f - f_\infty\|^{2-4\varepsilon}.$$

Step 3: Now a few complications will arise because we only have a control *from below* of $\mathcal{E}(f) - \mathcal{E}(f_\infty)$ in terms of $\|f - f_\infty\|$; so the fact that $\mathcal{E}(f) - \mathcal{E}(f_\infty)$ is of order E does not imply any lower bound on $\|f - f_\infty\|$, and then $\|f - f_\infty\|^{2-4\varepsilon}$ might be much, much higher than $\|f - f_\infty\|^{2+\varepsilon}$. To solve this difficulty, a small additional detour will be useful.

From Assumption 6(ii)-(iii) and interpolation,

(15.21) $\quad \|\Pi_1 f - f_\infty\|^{2+2\varepsilon} = \|\Pi_1 f - \Pi_1 f_\infty\|^{2+2\varepsilon} \leq C \|f - f_\infty\|^{2+\varepsilon};$

on the other hand,

(15.22) $\quad \|f - f_\infty\|^{2-4\varepsilon} \leq C \big(\|f - \Pi_1 f\|^{2-4\varepsilon} + \|\Pi_1 f - f_\infty\|^{2-4\varepsilon} \big).$

By using (15.21) and (15.22) in (15.19), and replacing the exponent $1 + \varepsilon/4$ by the worse exponent $1 + 2\varepsilon$ (which is allowed since $\mathcal{E}(f) - \mathcal{E}(\Pi_1 f) \leq \mathcal{E}(f) - \mathcal{E}(f_\infty)$ is uniformly bounded), we obtain

(15.23) $\quad -\widetilde{\mathcal{D}}(f) \leq -K[\mathcal{E}(f) - \mathcal{E}(\Pi_1 f)]^{1+2\varepsilon} - K\, a_{J-1} \|\Pi_1 f - f_\infty\|^{2+2\varepsilon}$
$$+ C \delta^{\frac{1}{1+\varepsilon}} \big(\|f - \Pi_1 f\|^{2-4\varepsilon} + \|\Pi_1 f - f_\infty\|^{2-4\varepsilon} \big),$$

where
$$\delta := \max_{0 \leq j \leq J-1} \frac{a_{j+1}^2}{a_j}.$$

Then from Assumption (7)(i)-(ii) (both the upper and the lower bounds are used in (ii)),

(15.24) $\quad -\widetilde{\mathcal{D}}(f) \leq -K[\mathcal{E}(f) - \mathcal{E}(\Pi_1 f)]^{1+2\varepsilon} - K\, a_{J-1} [\mathcal{E}(\Pi_1 f) - \mathcal{E}(f_\infty)]^{1+2\varepsilon}$
$$+ C \delta^{\frac{1}{1+\varepsilon}} [\mathcal{E}(f) - \mathcal{E}(\Pi_1 f)]^{1-3\varepsilon} + C \delta^{\frac{1}{1+\varepsilon}} [\mathcal{E}(\Pi_1 f) - \mathcal{E}(f_\infty)]^{1-3\varepsilon}.$$

Let us distinguish two cases:
First case: $\mathcal{E}(\Pi_1 f) - \mathcal{E}(f_\infty) \leq \mathcal{E}(f) - \mathcal{E}(\Pi_1 f)$.
Then
$$\mathcal{E}(f) - \mathcal{E}(f_\infty) \leq 2[\mathcal{E}(f) - \mathcal{E}(\Pi_1 f)],$$

and in particular

(15.25) $\quad \mathcal{E}(f) - \mathcal{E}(\Pi_1 f) \geq \dfrac{E}{4}.$

In that case we throw away the second negative term in (15.24), and bound the last term by the but-to-last one:

$$-\widetilde{\mathcal{D}}(f) \leq -K[\mathcal{E}(f) - \mathcal{E}(\Pi_1 f)]^{1+2\varepsilon} + C\delta^{\frac{1}{1+\varepsilon}}[\mathcal{E}(f) - \mathcal{E}(\Pi_1 f)]^{1-3\varepsilon}$$

$$(15.26) \qquad = -K\left(1 - \frac{C\delta^{\frac{1}{1+\varepsilon}}}{K}[\mathcal{E}(f) - \mathcal{E}(\Pi_1 f)]^{-5\varepsilon}\right)[\mathcal{E}(f) - \mathcal{E}(\Pi_1 f)]^{1+2\varepsilon}.$$

If

$$(15.27) \qquad \delta^{\frac{1}{1+\varepsilon}} \leq \frac{K}{2C}\left[\mathcal{E}(f) - \mathcal{E}(\Pi_1 f)\right]^{5\varepsilon}$$

(where K and C are the same constants as in (15.26)), then (15.26) can be bounded above by $-K'[\mathcal{E}(f) - \mathcal{E}(\Pi_1 f)]^{1+2\varepsilon}$, and by (15.25) this can also be bounded above by $-K''E^{1+2\varepsilon}$.

Finally, in view of (15.25) again, (15.27) is satisfied if

$$(15.28) \qquad \delta^{\frac{1}{1+\varepsilon}} \leq K'E^{5\varepsilon},$$

where $K' = 4^{-5\varepsilon}K/(2C)$. Since $\varepsilon \leq 1$ and E is uniformly bounded, a sufficient condition for (15.28) to hold is $\delta \leq K''E^{10\varepsilon}$.

Second case: $\mathcal{E}(\Pi_1 f) - \mathcal{E}(f_\infty) \geq \mathcal{E}(f) - \mathcal{E}(\Pi_1 f)$. In that case

$$(15.29) \qquad \mathcal{E}(\Pi_1 f) - \mathcal{E}(f_\infty) \geq \frac{E}{4},$$

and we retain from (15.24) that

$$-\widetilde{\mathcal{D}}(f) \leq -K\, a_{J-1}[\mathcal{E}(\Pi_1 f) - \mathcal{E}(f_\infty)]^{1+2\varepsilon} + C\delta^{\frac{1}{1+\varepsilon}}\left[\mathcal{E}(\Pi_1 f) - \mathcal{E}(f_\infty)\right]^{1-3\varepsilon};$$

and by a reasoning similar as the one above, this is bounded above by

$$-\frac{K}{2}\, a_{J-1}\left[\mathcal{E}(\Pi_1 f) - \mathcal{E}(f_\infty)\right]^{1+2\varepsilon}$$

as soon as

$$\delta^{\frac{1}{1+\varepsilon}} \leq K' a_{J-1} E^{5\varepsilon}.$$

This condition is fulfilled as soon as

$$\delta \leq K'' a_{J-1}^{1+\varepsilon} E^{10\varepsilon},$$

and, a fortiori (since $a_{J-1} \leq 1$) if

$$\delta \leq K''' a_{J-1}^{1+2\varepsilon} E^{10\varepsilon}.$$

In both cases, we have concluded that if $\varepsilon \leq \varepsilon_0$ and

$$\frac{a_{j+1}^2}{a_j} \leq K a_{J-1}^{1+2\varepsilon} E^{10\varepsilon},$$

then

$$-\widetilde{\mathcal{D}}(f) = \mathcal{L}'(f) \cdot (\mathcal{C}f - Bf) \leq -K' a_{J-1} E^{1+2\varepsilon}.$$

Up to the replacement of ε by $\varepsilon/2$ and ε_0 by $\varepsilon_0/2$, this is exactly the desired conclusion. □

PROOF OF THEOREM 51. Let $\overline{E}$ be such that $\mathcal{E}(f_0) - \mathcal{E}(f_\infty) \leq \overline{E}$. Since $\mathcal{E}(f(t))$ is a nonincreasing function of t,

$$\forall t \geq 0, \qquad \mathcal{E}(f(t)) - \mathcal{E}(f_\infty) \leq \overline{E}.$$

Let now $\varepsilon > 0$, and $E \in (0, \overline{E}]$. Let $[t_0, t_0 + T]$ be the time-interval where
$$\frac{E}{2} \leq \mathcal{E}(f(t)) - \mathcal{E}(f_\infty) \leq E;$$
this time-interval is well-defined (at least if T is a priori allowed to be infinite) since $\mathcal{E}(f(t)) - \mathcal{E}(f_\infty)$ is a continuous nonincreasing function. The goal is to show that if ε is small enough, then

(15.30) $$T \leq CE^{-\lambda\varepsilon},$$

where λ only depends on J, and C may depend on ε but not on E. When (15.30) is proven, it will follow from a classical argument that

(15.31) $$\mathcal{E}(f(t)) - \mathcal{E}(f_\infty) = O(t^{-1/((\lambda+1)\varepsilon)}).$$

Indeed, let $E_0 := \mathcal{E}(f_0) - \mathcal{E}(f_\infty)$; then $\mathcal{E}(f(t)) - \mathcal{E}(f_\infty)$ will be bounded by $E_0/2^{m+1}$ after a time
$$T_m := C\left(E_0^{-\lambda\varepsilon} + \left(\frac{E_0}{2}\right)^{-\lambda\varepsilon} + \left(\frac{E_0}{4}\right)^{-\lambda\varepsilon} + \ldots + \left(\frac{E_0}{2^m}\right)^{-\lambda\varepsilon}\right)$$
$$\leq C\left(\sum_{j=0}^{m} 2^{\lambda j \varepsilon}\right) E_0^{-\lambda\varepsilon} \leq C' 2^{\lambda m \varepsilon} E_0^{-\lambda\varepsilon}.$$

So $\mathcal{E}(f(t)) - \mathcal{E}(f_\infty) = O(2^{-(m+1)})$ after a time proportional to $2^{\lambda m \varepsilon}$, and (15.31) follows immediately.

Then $\mathcal{E}(f(t)) - \mathcal{E}(f_\infty) = O(t^{-\infty})$ since ε can be chosen arbitrarily small and λ does not depend on ε. From Assumption 7(i)-(ii),
$$\|f(t) - f_\infty\| \leq C[\mathcal{E}(f(t)) - E(f_\infty)]^{1/3}$$
(here $1/3$ could be $1/2 - \varepsilon$), so $\|f(t) - f_\infty\| = O(t^{-\infty})$ also. Finally, since $f(t)$ is bounded in all spaces X^s by Assumption 3, it follows by interpolation that $\|f(t) - f_\infty\|_s = O(t^{-\infty})$ for any $s > 0$.

So it all amounts to proving (15.30). Let K, K', k, ε_0 be provided by Theorem 52. (There is no loss of generality in taking the constants K appearing in (i) and (ii) to be equal.) Let then ε_1, K_1, ℓ be provided by Lemma A.23. For any $\varepsilon \leq \min(\varepsilon_0, \varepsilon_1)$ and any $t \in [t_0, t_0 + T]$ we have
$$\frac{E}{4} \leq \mathcal{L}(f(t)) \leq \frac{5E}{4};$$
$$\frac{d}{dt}\left[\mathcal{L}(f(t))\right] \leq -K' a_{J-1} E^{1+\varepsilon} \leq -K' K_1 E^{1+(\ell+1)\varepsilon}.$$
] So
$$E \geq \mathcal{L}(f(t_0)) - \mathcal{L}(f(t_0 + T)) \geq T K' K_1 E^{1+(\ell+1)\varepsilon},$$
and then (15.30) follows with $\lambda = \ell + 1$ (which eventually depends only on J). $\square$

16. Compressible Navier–Stokes system

In this section I start to show how to apply Theorem 51 on "concrete" examples. The compressible Navier–Stokes equations take the general form

(16.1) $$\begin{cases} \partial_t \rho + \nabla \cdot (\rho u) = 0 \\ \partial_t (\rho u) + \nabla \cdot (\rho u \otimes u) + \nabla p = \nabla \cdot \tau \\ \partial_t (\rho e) + \nabla \cdot (\rho e u + p u) = \nabla \cdot (\tau u) - \nabla \cdot q \end{cases}$$

where ρ is the density, u (vector-valued) is the velocity, e is the energy, q (vector-valued) is the heat flux, and τ (matrix-valued) is the viscous stress. In the case of perfect gases in dimension N, it is natural to use the following constitutive laws:

(16.2)
$$\begin{cases} p = \rho T \\ e = \dfrac{|u|^2}{2} + \dfrac{N}{2} T \\ \tau = 2\mu \{\nabla u\} \\ q = -\dfrac{N}{2} \kappa \nabla T, \end{cases}$$

where T is the temperature, μ is the viscosity, κ is the heat conductivity, and $\{\nabla u\}$ (matrix-valued) is the traceless symmetric strain:

$$\{\nabla u\}_{ij} = \frac{1}{2}\left(\frac{\partial u_i}{\partial x_j} + \frac{\partial u_j}{\partial x_i}\right) - \left(\frac{\nabla \cdot u}{N}\right)\delta_{ij},$$

and $\delta_{ij} = 1_{i=j}$. Then (16.1) takes the form

(16.3)
$$\begin{cases} \partial_t \rho + \nabla \cdot (\rho u) = 0; \\[4pt] \partial_t(\rho u) + \nabla \cdot (\rho u \otimes u) + \nabla(\rho T) = 2\mu \nabla \cdot \{\nabla u\}; \\[4pt] \partial_t\left(\rho \dfrac{|u|^2}{2} + \dfrac{N}{2} \rho T\right) + \nabla \cdot \left(\rho \dfrac{|u|^2}{2} u + \left(\dfrac{N+2}{2}\right) \rho u T\right) \\ \hspace{5cm} = 2\mu \nabla \cdot (u\{\nabla u\}) + \dfrac{N}{2}\kappa \Delta T. \end{cases}$$

(Note that

$$\nabla \cdot \{\nabla u\} = \mu \Delta u + \mu \left(1 - \frac{2}{N}\right)\nabla \nabla \cdot u,$$

so in the case considered here, the second Lamé coefficient is negative and equal to $-(2/N)\mu$, which is the borderline case.)

To avoid discussing boundary conditions I shall only consider the case when x varies in the torus $\mathbb{T}^N$. (Later on, for the Boltzmann equation we'll come to grips with boundary conditions a bit more.)

There are $N+2$ conservation laws for (16.1): total mass, total momentum (N scalar quantities) and total kinetic energy. Without loss of generality I shall assume

(16.4) $\qquad \int \rho = 1; \qquad \int \rho u = 0; \qquad \int \rho \dfrac{|u|^2}{2} + \dfrac{N}{2}\int \rho T = \dfrac{N}{2}.$

There is an obvious stationary state: $(\rho, u, T) \equiv (1, 0, 1)$. The goal of this section is the following conditional nonlinear stability result. The notation C^k stands for the usual space of functions whose derivatives up to order k are bounded.

THEOREM 55 (Conditional convergence for compressible Navier–Stokes). *Let* $t \to f(t) = (\rho(t), u(t), T(t))$ *be a C^∞ solution of (16.3), satisfying the uniform* bounds

(16.5) $$\begin{cases} \forall k \in \mathbb{N} \quad \sup_{t \geq 0} \Big(\|\rho(t)\|_{C^k} + \|u(t)\|_{C^k} + \|T(t)\|_{C^k} \Big) < +\infty; \\ \forall t \geq 0, \quad \rho(t) \geq \rho_m > 0; \quad T(t) \geq T_m > 0. \end{cases}$$

Then $\|f(t) - (1,0,1)\|_{C^k} = O(t^{-\infty})$ for all k.

PROOF OF THEOREM 55. Let us check that all assumptions of Theorem 51 are satisfied. Assumption 1 is satisfied with, say, $X^s = H^s(\mathbb{T}^N; \mathbb{R} \times \mathbb{R}^N \times \mathbb{R})$, where H^s stands for the usual L^2-Sobolev space of functions with s derivatives in L^2. To fulfill Assumption 2, define $C_s := \sup\{\|f(t)\|_{H^s}; \ t \geq 0\}$ and let

$$X = Y := \Big\{ f \in C^\infty(\mathbb{T}^N; \mathbb{R} \times \mathbb{R}^N \times \mathbb{R}); \quad \forall s, \quad \|f\|_s \leq C_s; \\ \rho \geq \rho_m; \quad T \geq T_m \Big\}.$$

(Note that necessarily $\rho_m \leq 1$, $T_m \leq 1$.)

To check Assumption 3, rewrite (16.3) in the nonconservative form

(16.6) $$\begin{cases} (\partial_t + u \cdot \nabla)\rho + \rho(\nabla \cdot u) = 0 \\ \\ (\partial_t + u \cdot \nabla)u + \nabla T + T\left(\dfrac{\nabla \rho}{\rho}\right) = \dfrac{2\mu}{\rho} \nabla \cdot \{\nabla u\} \\ \\ (\partial_t + u \cdot \nabla)T + \dfrac{2}{N} T(\nabla \cdot u) = \dfrac{4}{N} \dfrac{\mu}{\rho} |\{\nabla u\}|^2 + \dfrac{\kappa}{\rho} \Delta T, \end{cases}$$

and define
(16.7)
$$\mathcal{B}f = \Big(u \cdot \nabla \rho + \rho(\nabla \cdot u), \ u \cdot \nabla u + \nabla T + T\nabla(\log \rho), \ u \cdot \nabla T + \dfrac{2}{N} T(\nabla \cdot u) \Big);$$

(16.8) $$\mathcal{C}f = \Big(0, \ \dfrac{2\mu}{\rho} \nabla \cdot \{\nabla u\}, \ \dfrac{4}{N} \dfrac{\mu}{\rho} |\{\nabla u\}|^2 + \dfrac{\kappa}{\rho} \Delta T \Big).$$

Then (3) obviously holds true.

Assumption 4 is satisfied with $f_\infty = (1, 0, 1)$.

As usual in the theory of viscous compressible flows, an important difficulty to overcome is the fact that diffusion does not act on the ρ variable. So let $\Pi_1 = \Pi$ be defined by

$$\Pi(\rho, u, T) = (\rho, 0, 1).$$

Assumption 6 is obviously satisfied with this choice of nonlinear projection.

Next, let $\mathcal{E}$ be the negative of the usual entropy for perfect fluids:

$$\mathcal{E}(\rho, u, T) = \int \rho \log \rho - \dfrac{N}{2} \int \rho \log T.$$

Taking into account (16.4),

$$\mathcal{E}(\rho, u, T) - \mathcal{E}(1, 0, 1) = \int \rho \log \rho + \int \rho \dfrac{|u|^2}{2} + \dfrac{N}{2} \int \rho(T - \log T - 1);$$

$$\mathcal{E}(\rho, u, T) - \mathcal{E}(\Pi(\rho, u, T)) = \int \rho \dfrac{|u|^2}{2} + \dfrac{N}{2} \int \rho(T - \log T - 1).$$

Thanks to the uniform bounds from above and below on ρ and T, $\mathcal{E}(f) - \mathcal{E}(f_\infty)$ controls $\|f - f_\infty\|^2$ from above, and $\mathcal{E}(f) - \mathcal{E}(\Pi f)$ controls $\|f - \Pi f\|^2$ from above and below; so Assumption 7 is satisfied.

It only remains to check Assumption 8. By a classical computation, for any $f \in Y$,

$$\mathcal{E}'(f) \cdot (\mathcal{C}f) = -\left(2\mu \int \frac{|\{\nabla u\}|^2}{T} + \kappa \int \frac{|\nabla T|^2}{T^2}\right)$$

$$\leq -K \left(\int |\{\nabla u\}|^2 + \int \rho |\nabla T|^2\right),$$

where the last inequality follows again from the lower bound on T and the upper bound on ρ.

By Poincaré inequality, $\int \rho |\nabla T|^2$ controls $\int \rho (T - \langle T \rangle_\rho)^2$, where $\langle T \rangle_\rho = \int \rho T$ is the average of T with respect to ρ. In turn, this controls $\|T - 1\|^2 - 2(\langle T \rangle_\rho - 1)^2$. Since $\langle T \rangle_\rho - 1 = (-1/N) \int \rho |u|^2$, we conclude that there are positive constants K and C such that

$$\int \rho |\nabla T|^2 \geq K \|T - 1\|^2 - C \|u\|^2$$

for all $f \in Y$. On the other hand, by [**16**, Proposition 11],

$$\int |\{\nabla u\}|^2 \geq K' \|u\|^2.$$

All in all, there is a positive constant K such that

$$\int |\{\nabla u\}|^2 + \int \rho |\nabla T|^2 \geq K(\|T - 1\|^2 + \|u\|^2) = K \|f - \Pi f\|^2,$$

so Assumption 8(i) holds true.

By another classical computation, $\mathcal{E}'(f) \cdot (Bf) = 0$, so Assumption 8(ii) also holds true.

On the range of Π, the functional derivative Π' vanishes (because $\partial_t \rho = 0$ when $u = 0$), and $B(\rho, u, T) = (0, -\nabla \log \rho, 0)$. Then

$$(\mathrm{Id} - \Pi)'_{\Pi f} \cdot (B\Pi f) = B\Pi f = (0, -\nabla \log \rho, 0).$$

Thus

$$\left\|(\mathrm{Id} - \Pi)'_{\Pi f} \cdot (B\Pi f)\right\|^2 = \int |\nabla(\log \rho)|^2,$$

which under our assumptions controls $\int |\nabla \rho|^2$, and then by Poincaré inequality also $\|\rho - 1\|^2 = \|\Pi f - f_\infty\|^2$. This establishes Assumption 8(iii), and then the conclusion of the theorem follows from Theorem 51. □

17. Weakly self-consistent Vlasov–Fokker–Planck equation

One of the final goals of the theory which I have been trying to start in this memoir is the convergence to equilibrium for the nonlinear Vlasov–Poisson–Fokker–Planck equation with an external confinement. This kinetic model, of great importance in plasma physics, describes the evolution of a cloud of charged particles undergoing deterministic and random (white noise) forcing, friction, and influencing each other by means of Coulomb interaction.

Besides the fact that the regularity theory of the Vlasov–Poisson–Fokker–Planck equation is still at an early stage (to say the least), one meets serious difficulties when trying to apply Theorem 51 to this model, in particular because the problem is set in the whole space. So for the moment I shall be content to treat a simpler baby problem where (a) the confining potential is replaced by a periodic boundary condition; (b) the Coulomb interaction potential is replaced by a *small* and *smooth* potential. The smallness assumption is not only a technical simplification: It will prevent phase transition and guarantee the uniqueness of equilibrium state.

Even with these simplifications, the problem of convergence to equilibrium is nontrivial because the model is nonlinear and the diffusion only acts on the velocity variable. This will be a perfect example of application of Theorem 51.

Here the unknown $f = f(t, x, v)$ is a time-dependent probability density in phase space ($x \in \mathbb{T}^N$ stands for position and $v \in \mathbb{R}^N$ for velocity). The equation reads

(17.1)
$$\begin{cases} \partial_t f + v \cdot \nabla_x f + F[f](t,x) \cdot \nabla_v f = \Delta_v f + \nabla_v \cdot (fv) \\ F[f](t,x) = -\int \nabla W(x-y)\, f(t,y,w)\, dw\, dy. \end{cases}$$

Here $W \in C^\infty(\mathbb{T}^N)$ is even ($W(-z) = W(z)$), and without loss of generality $\int W = 0$. As we shall see later, if W is small enough in a suitable sense then the unique equilibrium for (17.1) is the Maxwellian with constant density:

$$f_\infty(x,v) = M(v) = \frac{e^{-\frac{|v|^2}{2}}}{(2\pi)^{N/2}}.$$

Since the total mass $\int f(t, x, v)\, dv\, dx$ is preserved with time, there is an a priori estimate on the force, like $\|F\|_{C^k} \leq C_k \|W\|_{C^k}$; so there is no real difficulty in adapting the proofs of regularity for the *linear* kinetic Fokker–Planck equation (see Appendix A.21). In this way one can establish the existence and uniqueness of a solution as soon as, say, the initial datum has finite moments of arbitrary order; and this solution will be smooth for positive times.

The goal of this section is to establish the following convergence result:

THEOREM 56 (Large-time behavior of the weakly self-consistent Vlasov–Fokker–Planck equation). *Let $W \in C^\infty(\mathbb{T}^N)$ satisfy $\int W = 0$. Let $f_0 = f_0(x,v)$ be a probability density on $\mathbb{T}^N \times \mathbb{R}^N$, such that $\int f_0(x,v)|v|^k\, dv\, dx < +\infty$ for all $k \in \mathbb{N}$, and let $f = f(t, x, v)$ be the unique smooth solution of (17.1). Let δ be so small that*

$$\delta + \frac{\delta^2 e^\delta}{2} < \frac{1}{2}.$$

If $\max |W| < \delta$ then

$$\|f(t, \cdot) - M\|_{L^1} = O(t^{-\infty}).$$

REMARK 57. It is not hard to show that the conclusion of Theorem 56 does not hold true without any size condition on W, since in general (17.1) can admit several stationary states. In the proof of Theorem 56 I shall show that there is only one stationary state as soon as $\max |W| < 1$; I don't know how good this bound is. The assumptions of the theorem are satisfied with $\delta = 0.38$, which does leave some margin of improvement.

PROOF OF THEOREM 56. The first step consists in establishing uniform regularity estimates; I shall only sketch them very briefly.

17. WEAKLY SELF-CONSISTENT VLASOV–FOKKER–PLANCK

First, one establishes differential inequalities on the "regularized" moments $M_k(t) = \int f(t,x,v)(1+|v|^2)^{k/2}\,dv\,dx$:

$$\frac{dM_k}{dt} \leq CM_{k-1} - KM_k,$$

where C and K are positive constants. (Here the fact that the position space is $\mathbb{T}^N$ induces a considerable simplifcation.) Then one deduces easily that each moment $\int f|v|^k$ remains bounded uniformly in time.

Next, by adapting the arguments in Appendix A.21, one can prove uniform Sobolev estimates of the form

$$\forall k \in \mathbb{N}, \quad \forall t_0 > 0, \quad \sup_{t \geq t_0} \|f(t,\cdot)\|_{H^k} < +\infty.$$

These bounds, combined with the moment estimates, imply the boundedness of the solution $f(t,\cdot)$ in all spaces X^s, where X^k is defined for $k \in \mathbb{N}$ by

(17.2) $$\|f\|_{X^k}^2 = \sum_{|\ell|+|m|\leq k} \int \left|\nabla_x^k \nabla_v^m f(x,v)\right|^2 (1+|v|^2)^k \, dv\, dx.$$

and X^s is defined by interpolation for noninteger s. It is easy to check that these spaces satisfy Assumption 1.

Next, hypoellipticity theory classically provides *local* strict positivity bounds on $f(t,x,v)$ for $t \geq t_0 > 0$. (An elementary method covering our needs is described in Appendix A.22; see Corollary A.21.) We deduce that $\rho(t,x) = \int f(t,x,v)\,dv$ is bounded below by a positive constant, uniformly in $t \geq t_0$. (Note: Once the local bounds are obtained, using a maximum principle method as in [**13**, Section 10] one can turn these bounds into global bounds $f(t,x,v) \geq K\,e^{-a|v|^2}$; but we do not need such a refinement.)

Up to changing the origin of time, we can now assume that f is uniformly bounded in all spaces X^s and that ρ satisfies a uniform lower bound. This determines a workspace

$$X = Y := \left\{f; \quad \forall s\ \|f\|_{X^s} \leq C_s;\ \ \rho \geq \rho_m > 0\right\},$$

as in Assumption 2.

Then we define

$$Bf = v \cdot \nabla_x f + F[f] \cdot \nabla_v f; \qquad \mathcal{C}f = \Delta_v f + \nabla_v \cdot (fv);$$

$$f_\infty = M(v); \qquad \Pi_1(f) = \Pi(f) = \rho M, \qquad \rho = \int f\,dv.$$

Assumptions 3, 4, 5 and 6 are readily checked.

Next let the free energy functional $\mathcal{E}$ be defined by

$$\mathcal{E}(f) = \int f \log f\,dv\,dx + \int f \frac{|v|^2}{2}\,dv\,dx + \frac{1}{2}\int \rho(x)\,\rho(y)\,W(x-y)\,dx\,dy.$$

By standard manipulations,

(17.3) $$\mathcal{E}(f) - \mathcal{E}(\Pi f) = \int_{\mathbb{T}^N \times \mathbb{R}^N} f \log \frac{f}{\rho M};$$

(17.4) $$\mathcal{E}(\Pi f) - \mathcal{E}(f_\infty) = \int_{\mathbb{T}^N} \rho \log \rho + \frac{1}{2}\int_{\mathbb{T}^N} \rho(x)\,\rho(y)\,W(x-y)\,dx\,dy.$$

The Csiszár–Kullback–Pinsker inequality implies the lower bound $\mathcal{E}(f)-\mathcal{E}(\Pi_1 f) \geq (1/2)\|f - \rho M\|_{L^1}^2$; then by interpolation of L^2 between L^1 and H^k (as in [**16**, Lemma 10]), one deduces

$$\mathcal{E}(f) - \mathcal{E}(\Pi f) \geq \frac{1}{2}\|f - \rho M\|_{L^1}^2 \geq K \|f - \rho M\|_{H^k}^{-\theta}\|f - \rho M\|_{L^2}^{2+\theta},$$

where θ is arbitrarily small if k is chosen large enough. This shows that Assumption 7(i) is satisfied.

On the other hand, since $\int W = 0$,

$$\mathcal{E}(\Pi f) - \mathcal{E}(f_\infty) = \int \rho \log \rho + \frac{1}{2} \int [\rho(x) - 1]\,[\rho(y) - 1]\, W(x - y)\, dx\, dy$$

$$\geq \frac{1}{2}\|\rho - 1\|_{L^1}^2 - \frac{1}{2}(\max |W|) \|\rho - 1\|_{L^1}^2.$$

By assumption, $\max |W| < 1$; so there is a constant $K > 0$ such that

$$\mathcal{E}(\Pi f) - \mathcal{E}(f_\infty) \geq K\|\rho - 1\|_{L^1}^2.$$

By interpolation again, this can be controlled from below by $\|\rho - 1\|_{L^2}^{2+\varepsilon}$ for arbitrarily small ε, and the left inequality in Assumption 7(ii) is satisfied. (This is the first time that we use the smallness assumption on W.) The right inequality in Assumption 7(ii) is easy.

By classical computations (see e.g. [**14**, Section 2]),

$$-\mathcal{E}'(f) \cdot (\mathcal{C}f) = \int f \left|\nabla_v \log \frac{f}{\rho M}\right|^2 dv\, dx \geq 2 \int f \log \frac{f}{\rho M} \geq \frac{1}{2}\|f - \rho M\|_{L^1}^2,$$

so there is no difficulty to establish Assumption 8(i). Assumption 8(ii) follows immediately since $\mathcal{E}'(f)\cdot(Bf) = 0$. So it only remains to establish Assumption 8(iii).

As in the example of the compressible Navier–Stokes system, the functional derivative Π' vanishes on the range of Π, so

$$(\mathrm{Id} - \Pi)'_{\Pi f} \cdot (B\Pi f) = B\Pi f = v \cdot \nabla_x(\rho M) - (\nabla W * \rho) \cdot \nabla_v(\rho M)$$
$$= v \cdot \nabla_x\bigl(\rho + \rho(W * \rho)\bigr)M.$$

Then

$$\left\|(\mathrm{Id} - \Pi)'_{\Pi f} \cdot (B\Pi f)\right\|^2 = \int \left|v \cdot \nabla_x\bigl(\rho + \rho(W * \rho)\bigr)\right|^2 M(v)\, dv\, dx$$
$$= \int \left|\nabla_x\bigl(\rho + \rho(\rho * W)\bigr)\right|^2 dx$$
$$\geq K \int \rho \left|\frac{\nabla \rho}{\rho} + \rho * W\right|^2 dx,$$

where the lower bound on ρ was used in the last inequality. Let

$$\mu(x) := \frac{e^{-(\rho * W)(x)}}{\int e^{-(\rho * W)}}.$$

Since μ is uniformly bounded from above and below, we can use a logarithmic Sobolev inequality with reference measure $\mu(x)\, dx$; so there is a positive constant

K such that

$$\int \rho \left|\frac{\nabla \rho}{\rho} + \rho * W\right|^2 dx = \int \rho \left|\nabla \log \frac{\rho}{\mu}\right|^2 dx$$

$$\geq K \int \rho \log \frac{\rho}{\mu} dx$$

(17.5)
$$= K \left(\int \rho \log \rho + \int \rho(\rho * W) - \log \int e^{-W*\rho}\right).$$

By assumption, $\max |W| \leq \delta$; so $|W * \rho| \leq \delta$, and

$$\left|e^{-W*\rho} - (1 - W * \rho)\right| \leq e^\delta \frac{(W*\rho)^2}{2} = e^\delta \frac{[W*(\rho-1)]^2}{2}$$

$$\leq \frac{e^\delta (\max |W|)^2}{2} \|\rho - 1\|_{L^1}^2 \leq \frac{\delta^2 e^\delta}{2} \|\rho - 1\|_{L^1}^2.$$

Since $\int (W * \rho) = 0$, it follows by integration of this bound that

$$\left|\int e^{-W*\rho} - 1\right| \leq \frac{\delta^2 e^\delta}{2} \|\rho - 1\|_{L^1}^2.$$

As a consequence,

$$\log \left(\int e^{-W*\rho}\right) \leq \frac{\delta^2 e^\delta}{2} \|\rho - 1\|_{L^1}^2.$$

From this bound and the inequality $|\int \rho(\rho * W)| \leq \delta \|\rho - 1\|_{L^1}^2$ again, we obtain

$$\int \rho \log \rho + \int \rho(\rho * W) - \log \int e^{-W*\rho}$$

$$\geq \frac{\|\rho - 1\|_{L^1}^2}{2} - \delta \|\rho - 1\|_{L^1}^2 - \frac{\delta^2 e^\delta}{2} \|\rho - 1\|_{L^1}^2$$

$$\geq \left(\frac{1}{2} - \delta - \frac{\delta^2 e^\delta}{2}\right) \|\rho - 1\|_{L^1}^2.$$

By assumption the coefficient in front of $\|\rho - 1\|_{L^1}^2$ is positive, and then we can use interpolation again to get

$$\int \rho \log \rho + \int \rho(\rho * W) - \log \int e^{-W*\rho} \geq K_\varepsilon \|\rho - 1\|_{L^2}^{2+\varepsilon} = K_\varepsilon \|\rho M - M\|_{L^2}^{2+\varepsilon}.$$

So Assumption 8(iii) holds. (Here again the smallness condition was crucially used.) Then all the assumptions of Theorem 51 are satisfied, and the conclusion follows at once. $\square$

18. Boltzmann equation

This last section is devoted to the Boltzmann equation; see [57] and the references therein for background and references on this model. I have personally devoted a considerable amount of research time on the problem of convergence to equilibrium for the Boltzmann equation, alone or in collaborations with Toscani and Desvillettes; a detailed account of this topic can be found in my lecture notes [56].

As in Section 17 the unknown is a time-dependent probability density $f = f(t, x, v)$ on the phase space. The variable x will be assumed to vary in a bounded

N-dimensional domain Ω_x, that will be either the torus $\mathbb{T}^N$, or a smooth bounded connected open subset of $\mathbb{R}^N$. The equation reads

$$(18.1) \quad \begin{cases} \dfrac{\partial f}{\partial t} + v \cdot \nabla_x f = Q(f,f) \\ \qquad = \displaystyle\int_{\mathbb{R}^N \times S^{N-1}} \Big[f(x,v')f(x,v'_*) - f(x,v)f(x,v_*) \Big] B(v-v_*,\sigma)\,d\sigma\,dv_* \\ v' = \dfrac{v+v_*}{2} + \dfrac{|v-v_*|}{2}\sigma; \qquad v'_* = \dfrac{v+v_*}{2} - \dfrac{|v-v_*|}{2}\sigma. \end{cases}$$

Here B is the collision kernel; for simplicity I shall restrict to the case $B = |v-v_*|$ (hard spheres interaction), but the analysis works as soon as Assumptions (5) and (19) in [16] are satisfied, which covers all physically relevant cases that I know of.

Three kinds of estimates play an important role in the modern theory of the Boltzmann equation: Sobolev estimates (in x and v variables), moment estimates and positivity estimates of the form $f \geq K_0 e^{-A_0 |v|^{q_0}}$. At least in some cases, the positivity estimates follow from regularity estimates [42], but I shall not address this issue here.

To continue the discussion it is necessary to take **boundary conditions** into account. I shall consider five cases: (i) periodic boundary conditions; (ii) bounce-back boundary conditions; (iii) specular reflection in a nonaxisymmetric domain; (iv) specular reflection in a spherically symmetric domain; (v) Maxwellian accommodation with constant wall temperature. Cases (i) to (iii) were already considered in [16], while cases (iv) and (v) are new and will be the occasion of interesting developments. Specular reflection in a general axisymmetric domain (not spherically symmetric) is intermediate between cases (iii) and (iv) and can probably be treated as a variant, but I have not tried to do so. Other conditions could be treated as a variant of (v), such as more general accommodation kernels, but they do not seem to cause any substantial additional difficulty. On the other hand, the techniques presented here are helpless to treat accommodation with *variable* wall temperature, for which the collision operator does not vanish; I shall add a few words about this issue in the end of the section.

18.1. Periodic boundary conditions. In this subsection I shall consider the Boltzmann equation (18.1) in the position space $\Omega_x = \mathbb{T}^N$ (the N-dimensional torus). Then there are $N+2$ conservation laws: total mass, total momentum (N components) and total kinetic energy. Without loss of generality, I shall assume

$$(18.2) \quad \int f\,dv\,dx = 1; \qquad \int f v\,dv\,dx = 0; \qquad \int f|v|^2\,dv\,dx = N.$$

Then the equilibrium state takes the form

$$f_\infty(x,v) = M(v) = \frac{e^{-\frac{|v|^2}{2}}}{(2\pi)^{N/2}}.$$

Our goal is the next convergence theorem:

THEOREM 58 (Convergence to equilibrium for the Boltzmann equation with periodic boundary conditions). *Let f be a solution of* (18.1) *in the spatial domain*

$\Omega_x = \mathbb{T}^N$, satisfying the conservation laws (18.2), and the uniform regularity estimates

(18.3)
$$\begin{cases} \forall s \geq 0 & \sup_{t \geq 0} \|f(t,\cdot)\|_{H^s(\Omega_x \times \mathbb{R}_v^N)} < +\infty; \\ \forall k \geq 0 & \sup_{t \geq 0} \int f(t,x,v)\,|v|^k\,dv\,dx < +\infty; \\ \forall (t,x,v) \in \mathbb{R}_+ \times \Omega_x \times \mathbb{R}_v^N, & f(t,x,v) \geq K_0\,e^{-A_0|v|^{q_0}}. \end{cases}$$

Then
$$\forall s \geq 0, \qquad \|f(t,\cdot) - M\|_{H^s} = O(t^{-\infty}).$$

REMARK 59. This theorem is nonempty, in the sense that, when f_0 is very smooth and close to equilibrium in a suitable sense, then (18.3) holds true. See the discussion in [16] for more information.

PROOF OF THEOREM 58. Let f satisfy the assumptions of Theorem 58. Let $(X^s)_{s \geq 0}$ be the scale of weighted Sobolev spaces already defined in the treatment of the Vlasov–Fokker–Planck equation (recall equation (17.2)). It follows from the assumptions that $C_s := \sup_{t \geq 0} \|f\|_s$ is finite for all s. We shall work in the spaces

$$X := \left\{ f;\ \|f\|_s \leq C_s;\ f(x,v) \geq K_0\,e^{-A_0|v|^{q_0}} \right\},$$
$$Y := \left\{ f;\ \|f\|_s \leq C'_s;\ f(x,v) \geq K'_0\,e^{-A_0|v|^{q_0}} \right\},$$

where C'_s, K'_0 will be determined later on. Then Assumptions 1, 2 and 3 are obviously satisfied.

Define
$$Bf = v \cdot \nabla_x f; \qquad \mathcal{C}f = Q(f,f).$$

Then Assumption 4(i) is obviously true, Assumption 4(ii) is satisfied since B is linear continuous $X^{s+1} \to X^s$, and Assumption 4(iii) is a consequence of [16, eq. (78)].

Assumption 5 holds true with $f_\infty = M$; notice that both the transport and the collision part vanish on f_∞.

Next, if $f = f(x,v)$ is given, define
$$\rho = \int f\,dv; \qquad u = \frac{1}{\rho}\int f v\,dv; \qquad T = \frac{1}{N\rho}\int f|v-u|^2\,dv;$$

and
$$M_{\rho u T} = \frac{\rho(x)\,e^{-\frac{|v-u(x)|^2}{2T(x)}}}{[2\pi T(x)]^{N/2}}.$$

It is easy to derive uniform estimates of smoothness on ρ, u and T in terms of the estimates on f; and to derive similarly strict positivity estimates on ρ, T: see [16, Proposition 7].

Now we can introduce the nonlinear projection operators:
$$\Pi_1 f = M_{\rho u T}; \qquad \Pi_2 f = M_{\rho 0 1}; \qquad \Pi_3 f = M.$$

By adjusting the constants C'_s, K'_0, we can ensure that $\Pi_j(X) \subset Y$. Then the rest of Assumption 6 follows easily.

The natural Lyapunov functional in the present case is of course Boltzmann's H functional:
$$\mathcal{E}(f) = H(f) = \int_{\mathbb{T}_x^N \times \mathbb{R}_v^N} f \log f \, dv \, dx.$$
By standard computations, taking into account (18.2), we have

(18.4) $$\mathcal{E}(f) - \mathcal{E}(\Pi_1 f) = \int f \log \frac{f}{M_{\rho u T}};$$

(18.5) $$\mathcal{E}(\Pi_1 f) - \mathcal{E}(f_\infty) = \int \rho \log \rho + \int \rho \frac{|u|^2}{2} + \int \rho(T - \log T - 1).$$

To find a lower bound on (18.4), it suffices to use the Csiszár–Kullback–Pinsker inequality and interpolation, as we did previously for the Vlasov–Fokker–Planck equation (recall (17.3); or [**16**, eq. (47)$_1$]). Upper and lower bounds for (18.5) can be obtained as we did before for the compressible Navier–Stokes equations. So Assumption 7 is satisfied.

Now the crucial step consists in checking Assumption 8. By a classical computation,
$$-H'(f) \cdot (\mathcal{C}f) = \int D\bigl(f(x,\cdot)\bigr) \, dx,$$
where $D(f)$ is Boltzmann's dissipation of information:
$$D(f) = \frac{1}{4} \int \Bigl(f(v')f(v_*') - f(v)f(v_*)\Bigr) \Bigl(\log f(v')f(v_*') - \log f(v)f(v_*)\Bigr)$$
$$B(v - v_*, \sigma) \, d\sigma \, dv \, dv_*.$$

Known entropy production estimates from [**58**] make it possible to estimate $D(f)$ from below by $K_\varepsilon \bigl[H(f) - H(M_{\rho u T})\bigr]^{1+\varepsilon}$. (Such estimates go back to [**54**]; see also [**56**] for a detailed account on this problem.) Then Assumption 8(i) follows easily, as in [**16**, Corollary 5].

Assumption 8(ii) is an immediate consequence of Assumption 8(i), since $H'(f) \cdot (Bf) = 0$.

It remains to establish Assumption 8(iii). For this we use Remark 49. According to [**16**, eq. (69)], if $f = M_{\rho u T}^f$ at time $t = 0$ and f evolves according to $\partial_t f + v \cdot \nabla_x f = 0$, then
$$\frac{d^2}{dt^2}\bigg|_{t=0} \|f - M_{\rho u T}^f\| \geq K \left(\int_{\mathbb{T}^N} |\nabla T|^2 + \int_{\mathbb{T}^N} |\{\nabla u\}|^2 \right),$$
where, as in Section 16,
$$\{\nabla u\}_{ij} = \frac{1}{2} \left(\frac{\partial u_i}{\partial x_j} + \frac{\partial u_j}{\partial x_i} \right) - \frac{1}{N} (\nabla \cdot u) \, \delta_{ij},$$
and $\nabla \cdot u$ is the divergence of u. According to [**16**, Section IV.2], there are constants K_1, K_2, K_3 only depending on N, such that
$$\begin{cases} \displaystyle\int_{\mathbb{T}^N} |\nabla T|^2 \geq K_1 \|T - 1\|^2 - C\|u\|^2; \\ \displaystyle\int_{\mathbb{T}^N} |\{\nabla u\}|^2 \geq K_2 \|\nabla u\|^2 \geq K_3 \|u\|^2. \end{cases}$$

This implies
$$\|(\mathrm{Id} - \Pi_1)'_{\Pi_1 f} \cdot (B\Pi_1 f)\|_{L^2}^2 \geq K(\|T-1\|^2 + \|u\|^2)$$
$$\geq K'\|\Pi_1 f - \Pi_2 f\|^2.$$

Next, according to [**16**, eq. (71)], if $f = M_{\rho 0 1}$ at time $t = 0$ and f evolves according to $\partial_t f + v \cdot \nabla_x f = 0$, then
$$\left.\frac{d^2}{dt^2}\right|_{t=0} \|f - M^f_{\rho 0 1}\| \geq K \int_{\mathbb{T}^N} |\nabla \rho|^2.$$

Combining this with a Poincaré inequality (see again [**16**, Section IV.2]), we deduce that
$$\|(\mathrm{Id} - \Pi_2)'_{\Pi_2 f} \cdot (B\Pi_2 f)\|_{L^2}^2 \geq K\|\rho - 1\|^2 \geq K'\|\Pi_2 f - f_\infty\|_{L^2}^2.$$

(This is in fact as in Section 17, if we set $W = 0$.)

This concludes the verification of Assumption 8, and the result follows by an application of Theorem 51. □

REMARK 60. A comparison with the proof of the same result in [**16**] shows that the crucial functional inequalities are all the same; but there are essential simplifications in that (a) it suffices to do the computations for Maxwellian states ("local equilibrium" in the language of [**16**]); and especially (b) there is no longer need for the tricky analysis of the system of differential inequalities. More explicitly, Sections III.3, V and VI of [**16**] are shortcut by the use of Theorem 51.

18.2. Bounce-back condition. Now let Ω_x be a bounded smooth open subset of $\mathbb{R}^N$; up to rescaling units we may assume that $|\Omega_x| = 1$ (the Lebesgue measure of the domain is normalized). In this subsection the boundary condition is of *bounce-back* type:

(18.6) $$x \in \partial\Omega_x \implies f(x,v) = f(x,-v).$$

A consequence of (18.6) is that $u = 0$ on $\partial\Omega_x$ (the mean velocity vanishes on the boundary).

Now there are only 2 conservation laws: mass and energy. So, without loss of generality, I shall assume

(18.7) $$\int f \, dv \, dx = 1; \qquad \int f |v|^2 \, dv \, dx = N.$$

The equilibrium is again the steady Maxwellian,
$$f_\infty(x,v) = M(v) = \frac{e^{-\frac{|v|^2}{2}}}{(2\pi)^{N/2}}.$$

Here is the analogue of Theorem 58:

THEOREM 61 (Convergence to equilibrium for the Boltzmann equation with bounce-back boundary conditions). *Let f be a solution of (18.1) in a smooth bounded connected spatial domain Ω_x, satisfying bounce-back boundary conditions, the conservation laws (18.7), and the uniform regularity estimates (18.3). Then*
$$\forall s \geq 0, \qquad \|f(t,\cdot) - M\|_{H^s} = O(t^{-\infty}).$$

PROOF OF THEOREM 61. The proof is quite similar to the proof of Theorem 58, however the sequence of projection operators is different:

$$\Pi_1 f = M_{\rho u T}; \qquad \Pi_2 f = M_{\rho u \langle T \rangle}; \qquad \Pi_3 f = M_{\rho 0 1}; \qquad \Pi_4 f = M,$$

where $\langle T \rangle = \int \rho T$ is the average temperature. According to [16, eq. (70)-(71)] and a reasoning similar to the one in the proof of Theorem 58,

(18.8)
$$\begin{cases} \left\| (\mathrm{Id} - \Pi_1)'_{\Pi_1 f} \cdot (B \Pi_1 f) \right\|^2 \geq K \|\nabla T\|^2; \\ \left\| (\mathrm{Id} - \Pi_2)'_{\Pi_2 f} \cdot (B \Pi_2 f) \right\|^2 \geq K \|\nabla^{\mathrm{sym}} u\|^2; \\ \left\| (\mathrm{Id} - \Pi_3)'_{\Pi_3 f} \cdot (B \Pi_3 f) \right\|^2 \geq K \|\nabla \rho\|^2, \end{cases}$$

where $\nabla^{\mathrm{sym}} u$ is the symmetrized gradient of u, that is

$$\left(\nabla^{\mathrm{sym}} u \right)_{ij} = \frac{1}{2} \left(\frac{\partial u_i}{\partial x_j} + \frac{\partial u_j}{\partial x_i} \right).$$

By Poincaré inequalities,

$$\|\nabla T\|^2 \geq K \|T - \langle T \rangle\|^2; \qquad \|\nabla \rho\|^2 \geq K \|\rho - 1\|^2.$$

By the classical Korn inequality, and the Poincaré inequality again (componentwise),

$$\|\nabla^{\mathrm{sym}} u\|^2 \geq K \|\nabla u\|^2 \geq K' \|u\|^2.$$

These estimates imply $\|(\mathrm{Id} - \Pi_j)'_{\Pi_j f} \cdot (B \Pi_j f)\|^2 \geq K \|\Pi_j f - \Pi_{j+1} f\|^2$ for all $j \in \{1, 2, 3\}$, so Assumption 8(iii) is satisfied in the end. Then Theorem 51 applies. $\square$

18.3. Specular reflection in a nonaxisymmetric domain. In this subsection the bounce-back boundary condition is replaced by the *specular reflection* condition:

$$x \in \partial \Omega_x \implies f(x, v) = f(x, R_x v), \qquad R_x v = v - 2 \langle v, n \rangle n.$$

This condition is more degenerate and the shape of the domain will influence the form of the equilibrium. For the moment I shall assume that the domain is *nonaxisymmetric* in dimension $N = 3$. The notation is the same as in Subsection 18.2.

THEOREM 62 (Convergence to equilibrium for the Boltzmann equation with nonaxisymmetric specular conditions). *Let f be a solution of (18.1) in a smooth bounded connected nonaxisymmetric spatial domain $\Omega_x \subset \mathbb{R}^3$, satisfying specular boundary condition, the conservation laws (18.7), and the uniform regularity estimates (18.3). Then*

$$\forall s \geq 0, \qquad \left\| f(t, \cdot) - M \right\|_{H^s} = O(t^{-\infty}).$$

PROOF OF THEOREM 62. The proof is entirely similar to the proof of Theorem 61, except that the condition $u = 0$ on the boundary is replaced by the weaker condition $u \cdot n = 0$, where n is the inner unit normal to Ω_x. Then the classical Korn inequality should be replaced by the Korn inequality established by Desvillettes and myself in [15]. $\square$

18.4. Specular reflection in a spherically symmetric domain.

In this subsection Ω_x is a bounded smooth connected spherically symmetric domain in $\mathbb{R}^3$; so, up to translation, Ω_x is either a ball ($|x| < R$) or a shell ($0 < r < |x| < R$). Again I shall assume that $|\Omega_x| = 1$. I shall write $N = 3$ to keep track of the role of the dimension in various formulas (certainly the analysis can be extended to more general domains, but one has to be careful about the meaning of the conservation of angular momentum).

Now there are $N + 2$ conservation laws: mass, kinetic energy and *angular momentum* (N scalar quantities). Without loss of generality, I shall assume

$$(18.9) \quad \int f\, dv\, dx = 1; \quad \int f(x,v)|v|^2\, dv\, dx = N;$$

$$\int f(x,v)(v \wedge x)\, dv\, dx = \mathbf{M} \in \mathbb{R}^N.$$

The existence of an equilibrium is not trivial if $\mathbf{M} \neq 0$, and the equilibrium does not seem to be explicit. It is a local Maxwellian with uniform temperature θ, but nonzero velocity u_∞ and nonhomogeneous density ρ_∞. The equations determining this equilibrium were studied, at the beginning of the nineties, by Desvillettes [12]. Here I shall suggest a variational approach to this problem, by means of the following lemma from elementary calculus of variations (the proof of which will be only sketched):

LEMMA 63 (stationary solutions in a spherically symmetric domain). *Let Ω_x be a spherically symmetric domain in $\mathbb{R}^N$, $N = 3$, $|\Omega_x| = 1$. Whenever ρ is a nonnegative integrable density on Ω_x, and $m \in L^1(\Omega_x; \mathbb{R}^N)$, define*

$$F(\rho, m) = \int \rho \log \rho - \frac{N}{2} \log\left(1 - \frac{1}{N}\int \frac{|m|^2}{\rho}\right).$$

Then there is a unique $(\rho_\infty, m_\infty) \in C^\infty(\Omega_x; \mathbb{R}_+ \times \mathbb{R}^N)$ which minimizes the functional F under the constraints

$$(18.10) \quad \int \rho = 1; \quad \int m(x) \wedge x\, dx = \mathbf{M}.$$

Moreover, ρ is strictly positive; and there are an antisymmetric matrix Σ_∞ and positive constants θ_∞ and Z such that for all $x \in \Omega_x$,

$$\frac{m(x)}{\rho(x)} = \Sigma_\infty x; \quad \rho_\infty(x) = \frac{e^{\frac{|\Sigma_\infty x|^2}{2\theta_\infty}}}{Z}.$$

SKETCH OF PROOF OF LEMMA 63. Write

$$\theta = 1 - \frac{1}{N}\int \frac{|m|^2}{\rho},$$

then

$$F(\rho, m) = \int \rho \log \rho + \frac{N}{2}(1 - \theta) + \frac{N}{2}(\theta - \log \theta - 1)$$

$$= \int \rho \log \rho + \int \frac{|m|^2}{2\rho} + \Psi(\theta),$$

where $\Psi(\theta) = (N/2)(\theta - \log \theta - 1)$.

By a classical computation, $(\rho, m) \longmapsto |m|^2/\rho$ is convex, so θ is a concave function of (ρ, m). Moreover, θ remains in $(0, 1)$, and on that interval Ψ is a convex *decreasing* function of θ. It follows that $\Psi(\theta)$ is a strictly convex function of (ρ, m). So

$$F : (\rho, m) \longmapsto \int \rho \log \rho + \int \frac{|m|^2}{2\rho} + \Psi(\theta)$$

is a strictly convex function of (ρ, m). This conclusion does not change if F is restricted on the domain defined by the *linear* constraints (18.10); so F has at most one minimizer.

The Euler–Lagrange equations for the minimization of F read

(18.11)
$$\begin{cases} \log \rho - \dfrac{|m|^2}{2\theta \rho^2} = \lambda_0; \\ \dfrac{N}{\theta}\left(\dfrac{m_i}{\rho}\right) = \varepsilon_{ijk} \lambda_j x_k, \end{cases}$$

where $(\lambda_j)_{0 \le j \le N}$ are constants, $(m_i)_{1 \le i \le N}$ are the components of m, and ε_{ijk} is defined by the equations $(a \wedge b)_i = \sum \varepsilon_{ijk} a_j b_k$. These equations imply that m/ρ is an antisymmetric linear function of x. In particular, the minimizer a priori lives in a finite-dimensional space. The rest of the lemma follows by classical arguments. □

The goal of the present subsection is the following result:

THEOREM 64 (Convergence to equilibrium for the Boltzmann equation with spherically symmetric specular conditions). *Let f be a solution of* (18.1) *in a smooth bounded connected spherically symmetric spatial domain $\Omega_x \subset \mathbb{R}^3$, satisfying specular boundary condition, the conservation laws* (18.9) *and the uniform regularity estimates* (18.3). *Then*

$$\forall s \ge 0, \qquad \|f(t, \cdot) - f_\infty\|_{H^s} = O(t^{-\infty}),$$

where

$$f_\infty(x, v) = \frac{e^{\frac{|\Sigma_\infty x|^2}{2\theta_\infty}}}{Z} \frac{e^{-\frac{|v - \Sigma_\infty x|^2}{2\theta_\infty}}}{(2\pi \theta_\infty)^{3/2}},$$

and the antisymmetric matrix Σ_∞, the positive constants Z and θ_∞ are provided by Lemma 63.

REMARK 65. The variable θ_∞ is the (uniform) equilibrium temperature; the velocity field in the stationary state is still rotating, and the density is lower near the interior of the box.

PROOF OF THEOREM 64. The only differences with the previously treated cases lie in the definition of the projection operators, and the verification of Assumptions 7(ii) and 8(iii).

In the present case, let

$$\Sigma := \langle \nabla^a u \rangle, \qquad \theta := \langle T \rangle_\rho;$$

more explicitly, Σ is the average value of the antisymmetric part of the matrix-valued field ∇u (the averaging measure is the normalized Lebesgue measure), while θ is the average value of the temperature (but now the averaging measure has density ρ). I shall identify the matrix Σ with the velocity field $x \longmapsto \Sigma x$, and θ

with the constant function $x \to \theta$. Then the sequence of projection operators is as follows:

$$\Pi_1 f = M_{\rho u T}; \qquad \Pi_2 f = M_{\rho u \theta}; \qquad \Pi_3 f = M_{\rho \Sigma \theta};$$
$$\Pi_4 f = f_\infty = M_{\rho_\infty \Sigma_\infty \theta_\infty}.$$

Once again the Lyapunov functional is

$$H(f) = \int f \log f.$$

After taking into account the conservation laws (18.9), one observes that

$$(18.12) \quad H(\Pi_1 f) - H(f_\infty) = \left(\int \rho \log \rho + \frac{N}{2} \int \rho(T - \log T - 1) + \int \rho \frac{|u|^2}{2} \right)$$
$$- \left(\int \rho_\infty \log \rho_\infty + \frac{N}{2} \int \rho_\infty (\theta - \log \theta - 1) + \int \rho_\infty \frac{|u_\infty|^2}{2} \right).$$

Let again $\Phi(\theta) = \theta - \log \theta - 1$: then by Jensen's inequality (in quantitative form),

$$\int \rho \Phi(T) \geq \Phi(\langle T \rangle_\rho) + K \|T - \langle T \rangle_\rho\|^2,$$

where K depends on the bounds on ρ and T. Plugging this in (18.12) and using the same notation as in Lemma 63, one obtains the lower bound

$$H(\Pi_1 f) - H(f_\infty)$$
$$= \frac{N}{2} \left(\int \rho \Phi(T) - \Phi(\langle T \rangle_\rho) \right) + \left[F(\rho, m) - F(\rho_\infty, m_\infty) \right]$$
$$\geq K \|T - \langle T \rangle_\rho\|^2 + \left[F(\rho, m) - F(\rho_\infty, m_\infty) \right].$$

The upper bound

$$H(\Pi_1 f) - H(f_\infty) \leq C \|T - \langle T \rangle_\rho\|^2 + \left[F(\rho, m) - F(\rho_\infty, m_\infty) \right]$$

is obtained in a similar way.

So to prove Assumption 7(ii), it suffices to check that

$$K \|\Pi_1 f - f_\infty\|^2 \leq F(\rho, m) - F(\rho_\infty, m_\infty) \leq C \|\Pi_1 f - f_\infty\|^2;$$

or, which amounts to the same,

$$(18.13) \quad K \left\| (\rho, m) - (\rho_\infty, m_\infty) \right\|^2 \leq F(\rho, m) - F(\rho_\infty, m_\infty)$$
$$\leq C \left\| (\rho, m) - (\rho_\infty, m_\infty) \right\|^2.$$

The upper bound is obvious from the definition, the bounds on (ρ, m) (which follow from the bounds on f) and the bounds on (ρ_∞, m_∞). To prove the lower bound, it suffices to establish the uniform convexity of F. Let

$$f(\rho, m) = \rho \log \rho + \frac{|m|^2}{2\rho}.$$

The Hessian of f has matrix

$$\begin{pmatrix} \dfrac{I_N}{\rho} & -\dfrac{m}{\rho^2} \\ -\dfrac{m}{\rho^2} & \dfrac{|m|^2}{\rho^3} + \dfrac{1}{\rho} \end{pmatrix},$$

where I_N stands for the $N \times N$ identity matrix; under our assumptions on ρ, this Hessian matrix is uniformly positive, so f is uniformly convex, and the same is true of the functional $F : (\rho, m) \longmapsto \int f(\rho, m)\, dx + \Psi(\theta)$. This conclusion does not change when one imposes the linear constraints (18.10), and the lower bound in (18.13) follows.

The last crucial step in the proof consists in the verification of Assumption 8(iii). To start with, we have, as in the previous subsection,

$$(18.14) \qquad \left\| (\mathrm{Id} - \Pi_1)'_{\Pi_1 f} \cdot (B\Pi_1 f) \right\|^2 \geq K \int |\nabla T|^2 + \int |\{\nabla u\}|^2$$
$$\geq K' \|T - \langle T \rangle_\rho\|^2,$$

which controls $\|\Pi_1 f - \Pi_2 f\|^2$.

Next,

$$\left\| (\mathrm{Id} - \Pi_2)'_{\Pi_2 f} \cdot (B\Pi_2 f) \right\|^2 \geq K \int |\nabla^{\mathrm{sym}} u|^2 \geq K' \|\nabla u - \Sigma\|^2,$$

where the second inequality follows from a version of Korn's inequality [**15**, eq. (1)]. Note that $\Sigma = \nabla(\Sigma x)$ (to avoid confusions I shall now write Σx for the map $x \to \Sigma x$), so one can apply again a Poincaré inequality to obtain in the end

$$(18.15) \qquad \left\| (\mathrm{Id} - \Pi_2)'_{\Pi_2 f} \cdot (B\Pi_2 f) \right\|^2 \geq K \|u - \Sigma x\|^2,$$

which controls $\|\Pi_2 f - \Pi_3 f\|^2$.

The gain from Π_3 is the main novelty. As a consequence of [**16**, eq. (65)],

$$(18.16) \qquad \frac{1}{M_{\rho \Sigma \theta}} \Big(\partial_t M_{\rho \Sigma \theta} + v \cdot \nabla_x M_{\rho \Sigma \theta} \Big)$$
$$= \left(\frac{\partial_t \rho + \Sigma \cdot \nabla \rho}{\rho} - \frac{N}{2} \frac{\partial_t \theta}{\theta} \right)$$
$$+ \left(\frac{v - u}{\sqrt{\theta}} \right) \cdot \left(\sqrt{\theta}\, \frac{\nabla \rho}{\rho} + \frac{\partial_t \Sigma + (\Sigma \cdot \nabla)\Sigma}{\sqrt{\theta}} \right)$$
$$+ \sum_i \left(\frac{v_i - u_i}{\sqrt{\theta}} \right)^2 \frac{\partial_t \theta}{\theta}.$$

The first and third lines do not bring any new estimate, so we focus on the second line. First note that

$$\big((\Sigma \cdot \nabla) \Sigma \big)_i (x) = \sum_{jk\ell} (\Sigma_{jk} x_k)\, \partial_j (\Sigma_{i\ell} x_\ell) = (\Sigma^2 x)_i.$$

(Do not mistake the symbol of summation with the matrix Σ.) Next, the equation for the mean velocity field u is $\partial_t u + u \cdot \nabla u + \nabla T + T\nabla(\log \rho) + (\nabla \cdot D)/\rho$, where D vanishes on the range of Π_1. Taking the antisymmetric part of this equation results in $\partial_t \nabla^a u = 0$, hence $\partial_t \Sigma = 0$. The conclusion is that $\partial_t \Sigma$ vanishes on the range

of Π_3. From all this information, we deduce that the second line of (18.16) can be simplified into
$$(v-u)\cdot\left(\frac{\nabla\rho}{\rho}+\frac{\Sigma^2 x}{\theta}\right),$$
where again $\Sigma^2 x$ is a shorthand for the map $x \to \Sigma^2 x$. It follows that

(18.17) $$\left\|(\mathrm{Id}-\Pi_3)'_{\Pi_3 f}\cdot(B\Pi_3 f)\right\|^2 \geq K\left\|\frac{\nabla\rho}{\rho}+\frac{\Sigma^2 x}{\theta}\right\|^2.$$

I shall now show that

(18.18) $$\left\|u-\Sigma x\right\|^2+\left\|\frac{\nabla\rho}{\rho}+\frac{\Sigma^2 x}{\theta}\right\|^2 \geq K\left(\|\rho-\rho_\infty\|^2+|\theta-\theta_\infty|^2\right.$$
$$\left.+\|\Sigma x-\Sigma_\infty x\|^2\right).$$

Since the right-hand side controls $\|\Pi_3 f - \Pi_4 f\|^{2+\varepsilon}$, in view of (18.17) and (18.15) this will imply

(18.19) $$\left\|(\mathrm{Id}-\Pi_2)'_{\Pi_2 f}\cdot(B\Pi_2 f)\right\|^2 + \left\|(\mathrm{Id}-\Pi_3)'_{\Pi_3 f}\cdot(B\Pi_3 f)\right\|^2 \geq K\|\Pi_3 f - \Pi_4 f\|^2,$$

completing the verification of Assumption 8(iii).

Since $|\theta-\theta_\infty| \leq C(\|\rho-\rho_\infty\|+\|u-u_\infty\|)$ and $\|\Sigma x - \Sigma_\infty x\| \leq C\|\Sigma - \Sigma_\infty\| \leq C'\|u-u_\infty\|$, to establish (18.18) it is sufficient to prove

(18.20) $$\left\|u-\Sigma x\right\|^2+\left\|\frac{\nabla\rho}{\rho}+\frac{\Sigma^2 x}{\theta}\right\|^2 \geq K\left(\|\rho-\rho_\infty\|^2+\|u-u_\infty\|^2\right).$$

In view of the bounds on ρ and u, and the uniform convexity of F (used above to check Assumption 7(ii)), inequality (18.20) will be a consequence of

(18.21) $$\left\|\nabla u-\Sigma\right\|^2+\left\|\frac{\nabla\rho}{\rho}+\frac{\Sigma^2 x}{\theta}\right\|^2 \geq K\left[F(\rho,m)-F(\rho_\infty,m_\infty)\right],$$

which we shall now check.

The following lemma will be useful:

LEMMA 66. *Let Φ be a K-uniformly convex function, defined and differentiable on a convex open subset of a Hilbert space $\mathcal{H}$, and let $\Lambda : \mathcal{H} \to \mathbb{R}^d$ be a linear map. If X_∞ minimizes Φ under the constraints $\Lambda(X)=c$, then*
$$\Phi(X)-\Phi(X_\infty) \leq (2K)^{-1} \inf_{\lambda\in(\mathrm{Ker}\,\Lambda)^\perp}\left\|\mathrm{grad}\,\Phi(X)+\lambda\right\|^2.$$

Postponing the proof of Lemma 66 for the moment, let us apply it to the uniformly convex functional
$$\Phi(\rho,m) = \int \rho\log\rho - \frac{N}{2}\log\left(1-\frac{1}{N}\int\frac{|m|^2}{\rho}\right)$$
and the linear map
$$\Lambda(\rho,m) = \left(\int\rho,\int m\wedge x\right)\in\mathbb{R}\times\mathbb{R}^N.$$
Then
$$\mathrm{grad}\,\Phi = \left(\log\rho-\frac{|m|^2}{2\theta\rho^2},\,\frac{m}{\theta\rho}\right),$$

and $(\operatorname{Ker}\Lambda)^\perp$ is made of vectors $\lambda = (\lambda_0, A)$, where $\lambda_0 \in \mathbb{R}$ and A is a (constant!) antisymmetric matrix. So Lemma 66 implies

$$(18.22) \quad F(\rho, m) - F(\rho_\infty, m_\infty) \leq C \inf_{\lambda_0 \in \mathbb{R};\; A^* = -A} \left(\left\| \log \rho - \frac{|m|^2}{2\theta \rho^2} \right\|^2 + \left\| \frac{m}{\theta \rho} - Ax \right\|^2 \right),$$

By Poincaré inequality, the bounds on θ, and Korn inequality,

$$(18.23) \quad \inf_{A^* = -A} \left\| \frac{m}{\theta \rho} - Ax \right\|^2 \leq C \inf_{A^* = -A} \left\| \nabla(u - Ax) \right\|^2 = C \left\| \nabla u - \langle \nabla^a u \rangle \right\|^2.$$

On the other hand,

$$\inf_{\lambda_0 \in \mathbb{R}} \left\| \log \rho - \frac{|m|^2}{2\theta \rho^2} \right\|^2 \leq 2 \inf_{\lambda_0} \left\| \log \rho - \frac{|\Sigma x|^2}{2\theta} - \lambda_0 \right\|^2 + 2 \left\| \frac{|\Sigma x|^2}{2\theta} - \frac{|m|^2}{2\theta \rho^2} \right\|^2$$

$$(18.24) \qquad \leq 2 \inf_{\lambda_0} \left\| \log \rho - \frac{|\Sigma x|^2}{2\theta} - \lambda_0 \right\|^2 + C \left\| u - \Sigma x \right\|^2.$$

By Poincaré inequality, the second term in the right-hand side of (18.24) can be bounded by a constant multiple of $\|\nabla u - \Sigma\|^2$. As for the first term, it can also be bounded by means of a Poincaré inequality:

$$\inf_{\lambda_0} \left\| \log \rho - \frac{|\Sigma x|^2}{2\theta} - \lambda_0 \right\|^2 = \left\| \log \rho - \frac{|\Sigma x|^2}{2\theta} - \left\langle \log \rho - \frac{|\Sigma x|^2}{2\theta} \right\rangle \right\|^2$$

$$\leq C \left\| \nabla \left(\log \rho - \frac{|\Sigma x|^2}{2\theta} \right) \right\|^2$$

$$= C \left\| \frac{\nabla \rho}{\rho} + \frac{\Sigma^2 x}{\theta} \right\|^2.$$

All in all,

$$\inf_{\lambda_0 \in \mathbb{R}} \left\| \log \rho - \frac{|m|^2}{2\theta \rho^2} \right\|^2 \leq C \left\| u - \Sigma x \right\|^2 + \left\| \frac{\nabla \rho}{\rho} + \frac{\Sigma^2 x}{\theta} \right\|^2.$$

This combined with (18.22) and (18.23) concludes the proof of (18.21). Then we can apply Theorem 51 and get the conclusion of Theorem 64. $\square$

PROOF OF LEMMA 66. Let $\widetilde{\Phi} : X \to \Phi(X_\infty + X)$. By assumption, 0 is a minimizer of $\widetilde{\Phi}$ on $\operatorname{Ker} \Lambda$. Since Φ is K-convex and differentiable, the same is true of $\widetilde{\Phi}$, so that

$$\widetilde{\Phi}(0) \geq \widetilde{\Phi}(X) - \left\langle \operatorname{grad}' \widetilde{\Phi}(X), X \right\rangle + \frac{K}{2} \|X\|^2,$$

where grad' stands for the gradient in the space $\operatorname{Ker} \Lambda$. It follows by Young's inequality that

$$\widetilde{\Phi}(X) - \widetilde{\Phi}(0) \leq \frac{\left\| \operatorname{grad}' \widetilde{\Phi}(X) \right\|^2}{2K}.$$

But $\operatorname{grad}' \widetilde{\Phi}$ is nothing but the orthogonal projection of $\operatorname{grad} \Phi(X)$ (in $\mathcal{H}$) onto $\operatorname{Ker} \Lambda$; so

$$\left\| \operatorname{grad}' \widetilde{\Phi}(X) \right\| = \inf_{\lambda \in (\operatorname{Ker} \Lambda)^\perp} \left\| \operatorname{grad} \widetilde{\Phi}(X) + \lambda \right\|.$$

The conclusion of Lemma 66 follows easily. $\square$

REMARK 67. I don't know if the term in Π_2 can be dispended with in (18.19); in any case this is an example where it is convenient to have the general formulation of Assumption 8(iii), rather than the simplified inequality (14.9). In the next subsection, another example will be presented where this possibility is crucially used (see Remark 69).

18.5. Maxwellian accommodation. In this subsection Ω_x will again be a bounded smooth connected open subset of $\mathbb{R}^N$ with unit Lebesgue measure, but now the boundary condition will be the Maxwellian accommodation with a *fixed* temperature T_w. Explicitly,

$$(18.25) \qquad x \in \partial\Omega_x \Longrightarrow f^+(x,v) = \left(\int f^-(x,v')\,|v'\cdot n|\,dv'\right) M_w(v),$$

where f^+ (resp. f^-) stands for the restriction of f to $\{v\cdot n > 0\}$ (resp. $\{v\cdot n < 0\}$), n is the *inner* unit normal vector, and M_w is a fixed "wall" Maxwellian:

$$M_w(v) = \frac{e^{-\frac{|v|^2}{2T_w}}}{(2\pi)^{\frac{N-1}{2}} T_w^{\frac{N+1}{2}}}.$$

(The analysis would go through if one would impose a more general condition involving a reflection kernel $C(v' \to v)$, as in [**11**, Chapter 1].) An important identity which follows from (18.25) is

$$(18.26) \qquad \forall x \in \partial\Omega_x, \quad \int_{\mathbb{R}^N} f(x,v)\,(v\cdot n) = 0;$$

equivalently, the mean velocity satisfies

$$(18.27) \qquad \forall x \in \partial\Omega_x, \quad u\cdot n = 0.$$

In this case there is only one conservation, namely the total mass. Without loss of of generality, I shall assume that the solution is normalized so that

$$(18.28) \qquad \int f\,dv\,dx = 1.$$

Then the unique equilibrium is the Maxwellian distribution with constant temperature equal to the wall temperature:

$$(18.29) \qquad f_\infty(x,v) = \frac{e^{-\frac{|v|^2}{2T_w}}}{(2\pi)^{N/2}}.$$

THEOREM 68 (Convergence to equilibrium for the Boltzmann equation with Maxwellian accommodation). *Let f be a solution of (18.1) in a smooth bounded connected spatial domain $\Omega_x \subset \mathbb{R}^N$ with $|\Omega_x| = 1$. Assume that f satisfies the boundary condition (18.25), the conservation laws (18.28) and the uniform regularity estimates (18.3). Then*

$$\forall s \geq 0, \qquad \|f(t,\cdot) - f_\infty\|_{H^s} = O(t^{-\infty}),$$

where f_∞ is defined by (18.29).

PROOF OF THEOREM 68. The proof follows again the same pattern as in all the previous theorems in this section. However, the Lyapunov functional is not Boltzmann's H functional, but a modified version of it:

$$\mathcal{E}(f) = \int f \log f + \frac{1}{2T_w} \int f|v|^2\,dv\,dx.$$

Moreover, the sequence of nonlinear projection operators will be

$$\Pi_1 f = M_{\rho u T}; \quad \Pi_2 f = M_{\rho u T_w}; \quad \Pi_3 f = M_{\rho 0 T_w}; \quad \Pi_4 f = M_{1 0 T_w}.$$

In particular,

$$\mathcal{E}(\Pi_1 f) - \mathcal{E}(f_\infty)$$
$$= \int \rho \log \rho - \frac{N}{2} \int \rho \log T + \frac{1}{T_w} \left(\int \rho \frac{|u|^2}{2} + \frac{N}{2} \int \rho T \right)$$
$$- \frac{N}{2} \log T_w + \frac{N}{2}$$
$$= \int \rho \log \rho + \frac{1}{T_w} \int \rho \frac{|u|^2}{2} + \frac{N}{2} \int \rho \left(\frac{T}{T_w} - \log \frac{T}{T_w} - 1 \right).$$

From this it is easy to check Assumption 7(ii).

The interesting features of this case reveal themselves when we try to check Assumption 8. First, by a classical computation,

$$\mathcal{D}(f) = -\mathcal{E}'(f) \cdot (Bf) = \int_{\Omega_x} D(f(x, \cdot)) \, dx + \int_{\Omega_x \times \mathbb{R}^N} (\log f + 1)(-v \cdot \nabla_x f)$$
$$+ \frac{1}{2T_w} \int_{\Omega_x \times \mathbb{R}^N} (v \cdot \nabla_x f) |v|^2 \, dv \, dx$$
$$= \int_{\Omega_x} D(f(x, \cdot)) \, dx - \int_{\partial \Omega_x \times \mathbb{R}^N} \left(f \log f(x, v) + f(x, v) \frac{|v|^2}{2T_w} \right) |v \cdot n| \, dv \, dx$$
$$= \int_{\Omega_x} D(f(x, \cdot)) \, dx + \int_{\partial \Omega_x \times \mathbb{R}^N} f \log \frac{f}{e^{-\frac{|v|^2}{2T_w}}} (v \cdot n) \, dv \, dx.$$

As before, he first term in the right-hand side is controlled below by $K[\mathcal{E}(f) - \mathcal{E}(\Pi_1 f)]^{1+\varepsilon}$. The second term needs some rewriting. In view of (18.26) and (18.25), we have, writing $\rho^-(x) = \int f^-(x,v) |v \cdot n| \, dv$,

$$\int_{\partial \Omega_x \times \mathbb{R}^N} -f \log \frac{f}{e^{-\frac{|v|^2}{2T_w}}} (v \cdot n) \, dv \, dx = -\int f \log \left(\frac{f}{\rho^- M_w} \right) (v \cdot n) \, dv \, dx$$
$$= \int_{v \cdot n < 0} f \log \left(\frac{f}{\rho^- M_w} \right) |v \cdot n| \, dv \, dx$$
$$= \int_{v \cdot n < 0} f |v \cdot n| \log \left(\frac{f |v \cdot n|}{\rho^- M_w |v \cdot n|} \right) dv \, dx.$$

This quantity takes the form of a nonnegative information functional, as a particular case of the Darrozès–Guiraud–Cercignani inequality [**11**, Chapter 1]. The Csiszár–Kullback–Pinsker inequality will give an explicit lower bound: For each $x \in \partial \Omega_x$,

$$\int_{v \cdot n < 0} f |v \cdot n| \log \left(\frac{f |v \cdot n|}{\rho^- M_w |v \cdot n|} \right) dv$$
$$\geq \frac{1}{2\rho^-(x)} \left\| f |v \cdot n| - \rho^- M_w |v \cdot n| \right\|^2_{L^1(\{v \cdot n < 0\}; |v \cdot n| \, dv)}$$
$$= \frac{1}{2\rho^-(x)} \left\| f |v \cdot n| - \rho^- M_w |v \cdot n| \right\|^2_{L^1(|v \cdot n| \, dv)}.$$

18. BOLTZMANN EQUATION

After interpolation and use of smoothness bounds, we conclude that
$$\mathcal{D}(f) \geq K\big[\mathcal{E}(f) - \mathcal{E}(\Pi_1 f)\big]^{1+\varepsilon} + K\|f - \rho^- M_w\|^{2+\varepsilon}_{L^q(\partial\Omega_x \times \mathbb{R}^N; \, |v \cdot n|\,(1+|v|^2)^{q/2}\,dx\,dv)}, \tag{18.30}$$
where q is arbitrarily large and ε is arbitrarily small. A useful consequence of (18.30) is
$$\mathcal{D}(f) \geq K\|T - T_w\|^{2+\varepsilon}_{L^q(\partial\Omega_x)} + K\|u\|^{2+\varepsilon}_{L^q(\partial\Omega_x)}, \tag{18.31}$$
where again q is arbitrarily large.

The other estimates are similar to the ones in the previous subsections:
$$\begin{cases} \left\|(\mathrm{Id} - \Pi'_1)_{\Pi_1 f} \cdot (B\Pi_1 f)\right\|^2 \geq K\big(\|\nabla T\|^2 + \|\{\nabla u\}\|^2\big); \\[4pt] \left\|(\mathrm{Id} - \Pi'_2)_{\Pi_2 f} \cdot (B\Pi_2 f)\right\|^2 \geq K\|\nabla^{\mathrm{sym}} u\|^2; \\[4pt] \left\|(\mathrm{Id} - \Pi'_3)_{\Pi_3 f} \cdot (B\Pi_3 f)\right\|^2 \geq K\|\nabla \rho\|^2. \end{cases} \tag{18.32}$$

Thanks to (18.31) and (18.32)$_1$,
$$\mathcal{D}(f) + \left\|(\mathrm{Id} - \Pi'_1)_{\Pi_1 f} \cdot (B\Pi_1 f)\right\|^2 \geq K\Big(\|\nabla T\|^2_{L^2(\Omega_x)} + \|T - T_w\|^{2+\varepsilon}_{L^q(\partial\Omega_x)}\Big)$$
$$\geq K'\|T - T_w\|^{2+\varepsilon}_{L^p(\Omega_x)},$$
where the latter inequality comes from, say, the trace Sobolev inequality if, say, $p = (2N)/(N-2)$ and $q = 2(N-1)/(N-2)$. (If $N = 2$ a slightly different argument based on a variant of the Moser–Trudinger inequality can be used to give the same result.) After interpolation one concludes that
$$\mathcal{D}(f) + \left\|(\mathrm{Id} - \Pi'_1)_{\Pi_1 f} \cdot (B\Pi_1 f)\right\|^2 \geq K\|T - T_w\|^{2+\varepsilon} \geq K'\|\Pi_1 f - \Pi_2 f\|^{2+\varepsilon'}. \tag{18.33}$$

Next, if Ω_x is not axisymmetric, then the boundary condition (18.27), the Korn inequality from [15] and the Poincaré inequality imply
$$\left\|(\mathrm{Id} - \Pi'_2)_{\Pi_2 f} \cdot (B\Pi_2 f)\right\|^2 \geq K\|\nabla^{\mathrm{sym}} u\|^2 \tag{18.34}$$
$$\geq K'\|u\|^2 \geq K''\|\Pi_2 f - \Pi_3 f\|^{2+\varepsilon}.$$

If Ω_x is axisymmetric, the previous argument breaks down, but we can use (18.31) and replace (18.34) by

(18.35)
$$\mathcal{D}(f) + \left\|(\mathrm{Id} - \Pi'_2)_{\Pi_2 f} \cdot (B\Pi_2 f)\right\|^2 \geq K\|\nabla^{\mathrm{sym}} u\|^2 + \|u\|^{2+\varepsilon}_{L^2(\partial\Omega)}$$
$$\geq K'\|u\|^{2+\varepsilon}_{L^2(\partial\Omega)} \geq K''\|\Pi_2 f - \Pi_3 f\|^{2+\varepsilon'},$$
where the but-to-last inequality follows from a *trace Korn inequality* (Proposition A.27 in Appendix A.23).

Finally,
$$\left\|(\mathrm{Id} - \Pi'_3)_{\Pi_3 f} \cdot (B\Pi_3 f)\right\|^2 \geq K\|\nabla \rho\|^2$$
$$\geq K\|\rho - 1\|^2 \geq K'\|\Pi_3 f - \Pi_4 f\|^{2+\varepsilon'}.$$

Then Assumption 8(iii) is satisfied, and one can use Theorem 51 to prove Theorem 68. □

REMARK 69. This is an example where the range of Π_1 is much larger than the set where the dissipation $\mathcal{D}$ vanishes. Trying to devise a projection operator onto the space where $\mathcal{D}$ vanishes gives rise to a horrendous nonlocal variational problem whose solution is totally unclear. On the other hand, inequalities (18.33) and (18.35) would be false without the contribution of $\mathcal{D}(f)$. In this example we see that the possibility to use the generalized condition appearing in Assumption 8(iii), rather than the simplified condition (14.9), leads to a great flexibility.

18.6. Further comments. In many important situations (variable wall temperature, evaporation problems, etc.), one is led to study *non-Maxwellian* stationary solutions of the Boltzmann equation; then there is usually no variational principle for these solutions, and the mere existence of stationary solutions is a highly nontrivial problem, see e.g. [**1, 2, 3**].

From the technical point of view, the non-Maxwellian nature of the stationary state means that if one defines $B = v \cdot \nabla_x$ (transport operator) and $\mathcal{C} = Q$ (collision operator), then the equations $Bf_\infty = 0$ and $\mathcal{C}f_\infty = 0$ cease to hold. No need to say, Theorem 51 collapses, and it is quite hard to figure out how to save it.

There is a thin analogy with the (linear) problem of the oscillator chain considered in Subsection 9.2 in the case when the two temperatures are not equal; in that case a change of reference measure, based on Proposition 5(ii), was at least able to reduce the problem to one of the type $A^*A + B$, $B^* = -B$. By analogy, one could imagine that a first step to come to grips with the quantitative analysis of stability for non-Maxwellian stationary solutions consists in re-defining the "antisymmetric" and the "diffusive" parts of the Boltzmann equation by performing some change of reference measure. Even this first step is nontrivial.

Appendices

This last part is devoted to some technical appendices used throughout the memoir, some of them with their own interest.

In Appendix A.19 I have gathered some sufficient conditions for a probability measure to admit a Poincaré inequality. After recalling some well-known criteria for Poincaré inequality in $\mathbb{R}^n$, I shall prove some useful results about tensor products; they might belong to folklore in certain mathematical circles, but I am not aware of any precise reference.

Appendices A.20, A.21 and A.22 are devoted to some properties of the linear (kinetic) Fokker–Planck equation. First in Appendix A.20 I shall prove a uniqueness theorem; the method is quite standard, although computations are a bit tricky.

Then in Appendix A.21 I shall present a new strategy to get hypoelliptic regularization estimates for the kinetic Fokker–Planck equation. This contribution is much more original and could be considered as a research paper on its own right. The method has the advantage to be very elementary, to avoid fractional derivatives as well as localization, and to yield optimal exponents of decay in short time. As Nash's theory of elliptic regularity, it is based on differential equations satisfied by certain functionals of the solutions. The results are nonstandard in several respects: they are global, directly yield pointwise in time estimates, and apply for initial data that do not lie in an L^2-type space.

A closely related, but somewhat simpler strategy was found independently and almost simultaneously by Frédéric Hérau. I shall explain his method in Subsection A.21.2, and develop it into an abstract theorem of global regularization applying to the same kind of operators that have been considered in Part I of this memoir. In Subsection A.21.4 I discuss further "entropic" estimates of hypoelliptic regularization, where regularity is quantified by the Fisher information.

In Appendix A.22 I shall present some nonoptimal, but elementary methods to establish Gaussian lower bounds for Fokker–Planck-type equations.

Finally, in Appendix A.23 I have gathered various technical lemmas and functional inequalities which are used throughout the memoir. I draw the attention of the reader to the "distorted Nash inequality" appearing in Lemma A.25, which might have an interesting role to play in the future for "global" hypoelliptic regularization estimates.

I warmly thank François Bouchut and Frédéric Hérau for many discussions about kinetic hypoellipticity. I developed the core of the method of Appendix A.21 during a stay in Reading University, from January to March 2003; thanks are due to Mike Cullen for his hospitality. The abstract regularization theorem from Subsection A.21.2 grew out from discussions with Denis Serre. I also acknowledge useful conversations with Michael Christ in Berkeley. Remarks A.16 and A.17 about pointwise estimates on the fundamental solution, and Appendix A.22 as well, were written during a stay in Kyoto University, in July and August 2008, long after the completion of the rest of these notes; it is a pleasure to thank Kazuo Aoki and the Department of Mechanical Engineering and Science in Kyoto for their hospitality, and the Japan Society for Promotion of Science for its support.

A.19. Some criteria for Poincaré inequalities

To begin with, I shall recall a popular and rather general criterion for Poincaré inequalities in $\mathbb{R}^n$.

THEOREM A.1. *Let $V \in C^2(\mathbb{R}^n)$, such that e^{-V} is a probability density on $\mathbb{R}^n$. If*

$$\text{(A.19.1)} \qquad \frac{|\nabla V(x)|^2}{2} - \Delta V(x) \xrightarrow[|x| \to \infty]{} +\infty,$$

then μ satisfies a Poincaré inequality.

PROOF. The starting point is a well-known estimate, see e.g. [**17**, Proof of Theorem 6.2.21]. Let $h \in C_c^1(\mathbb{R}^n)$, define $g = h\,e^{-aV}$, $a > 0$; then after expanding the square norm of the gradient and integrating by parts, one finds

$$\int |\nabla h|^2 \, e^{-V} = \int |\nabla g|^2 \, e^{(2a-1)V} + \int h^2 \left[(a - a^2)|\nabla V|^2 - a\Delta V\right] e^{-V}.$$

The choice $a = 1/2$ gives

$$\text{(A.19.2)} \qquad \int w h^2 \, d\mu \le 2 \int |\nabla h|^2 \, d\mu, \qquad w = \frac{|\nabla V|^2}{2} - \Delta V.$$

Let $R_0 > 0$ be large enough that $w(|x|) > 0$ for $|x| \ge R_0$. For $R > R_0$, define $\varepsilon(R) := [\inf\{w(|x|); |x| \ge R\}]^{-1}$; then $\varepsilon(R) \to 0$ as $R \to \infty$. So it follows from (A.19.2) that

$$\text{(A.19.3)} \qquad \int_{|x| \ge R} h^2 \, d\mu \le \varepsilon(R) \left[2 \int |\nabla h|^2 \, d\mu - (\inf w) \int h^2 \, d\mu\right].$$

Now let $h \in C^1(\mathbb{R}^n, \mathbb{R})$ with $\int h \, d\mu = 0$. For any $R > 0$, let B_R be the ball of radius R in $\mathbb{R}^n$, and let μ_R be the restriction of μ to B_R (normalized to be a probability measure). Since B_R is bounded, μ_R satisfies a Poincaré inequality with a constant $P(R)$ depending on R, so

$$\int h^2 \, d\mu_R \le P(R) \int |\nabla h|^2 \, d\mu_R + \left(\int h \, d\mu_R\right)^2.$$

Of course μ_R has density $(\mu[B_R])^{-1} e^{-V(x)} 1_{|x| \le R}$. If R is large enough, then $\mu[B_R] \ge 1/2$, so

$$\text{(A.19.4)} \qquad \int_{|x| \le R} h^2 \, e^{-V} \le P(R) \int_{|x| \le R} |\nabla h|^2 \, e^{-V} + 2 \left(\int_{|x| \le R} h \, e^{-V}\right)^2.$$

Since $\int h\,e^{-V} = 0$ and e^{-V} is a probability density,

$$\text{(A.19.5)} \qquad \left(\int_{|x| \le R} h\,e^{-V}\right)^2 = \left(\int_{|x| > R} h\,e^{-V}\right)^2 \le \int_{|x| > R} h^2 e^{-V}.$$

Plugging this into (A.19.4), one deduces that

$$\text{(A.19.6)} \qquad \int h^2 \, e^{-V} \le P(R) \int |\nabla h|^2 \, e^{-V} + 3 \int_{|x| > R} h^2 \, e^{-V}.$$

Combining this with (A.19.3), we recover

$$\int h^2 \, e^{-V} \le [P(R) + 6\varepsilon(R)] \int |\nabla h|^2 \, e^{-V} - 3(\inf w)\varepsilon(R) \int h^2 \, e^{-V}.$$

So, if R is large enough that $3(\inf w)\varepsilon(R) > -1$, one has
$$\int h^2 e^{-V} \leq \left(\frac{P(R) + 6\varepsilon(R)}{1 + 3(\inf w)\varepsilon(R)}\right) \int |\nabla h|^2 e^{-V}.$$
This concludes the proof of Theorem A.1. □

The sequel of this Appendix is devoted to Poincaré inequalities in product spaces. It is well-known that "spectral gap inequalities tensorize", in the following sense: If each L_ℓ ($\ell = 1, 2$) is a nonnegative operator on a Hilbert space $\mathcal{H}_\ell$, admitting a spectral gap κ_ℓ, then $L_1 \otimes I + I \otimes L_2$ admits a spectral gap $\kappa = \min(\kappa_1, \kappa_2)$. Now the goal is to extend this result in a form which allows multipliers. I shall start with an abstract theorem and then particularize it.

THEOREM A.2. *For $\ell = 1, 2$, let L_ℓ be a nonnegative unbounded operator on a Hilbert space $\mathcal{H}_\ell$, admitting a finite-dimensional kernel, and a spectral gap $\kappa_\ell > 0$. Let $\overline{M}$ be a nonnegative unbounded operator acting on $\mathcal{H}_2$, whose restriction to the kernel $\mathcal{K}_2$ of L_2 is bounded and coercive. Then the unbounded operator*
$$L = L_1 \otimes \overline{M} + I \otimes L_2$$
admits a spectral gap $\kappa > 0$. More precisely, for any nonnegative operator $M \leq \overline{M}$, such that the restriction $M|_{\mathcal{K}_2}$ of M to $\mathcal{K}_2$ satisfies $\lambda I \leq M \leq \Lambda I$, one has
$$\kappa \geq \min\left(\frac{\kappa_2}{2}, \frac{\kappa_2}{16}\frac{\lambda}{\Lambda^2}, \frac{\kappa_1}{2}\lambda\right).$$

THEOREM A.3. *(i) For $\ell = 1, 2$, let (X_ℓ, μ_ℓ) be a probability space, and let L_ℓ be a nonnegative operator on $\mathcal{H}_\ell = L^2(\mu_\ell)$, whose kernel is made of constant functions, admitting a spectral gap $\kappa_\ell > 0$. Let $\overline{m}$ be a nonnegative measurable function on X_2, which does not vanish μ_2-almost everywhere, and $\overline{M}$ be the multiplication operator by $\overline{m}$. Then the unbounded operator*
$$L = L_1 \otimes \overline{M} + I \otimes L_2$$
admits a spectral gap $\kappa > 0$. More precisely, for any nonnegative function $m \leq \overline{m}$, lying in $L^2(\mu_2)$,
$$\kappa \geq \min\left(\frac{\kappa_2}{2}, \frac{\kappa_2}{16}\frac{\|m\|_{L^1}^2}{\|m\|_{L^2}^2}, \frac{\kappa_1}{2}\|m\|_{L^1}\right).$$

(ii) More generally, for each $\ell \in \{1, \ldots, N\}$, let (X_ℓ, μ_ℓ) be a probability space, and let L_ℓ ($1 \leq \ell \leq N$) be a nonnegative symmetric operator on $\mathcal{H}_\ell = L^2(\mu_\ell)$, whose kernel is made of constant functions, admitting a spectral gap κ_ℓ; let $\overline{m}_\ell$ be a nonnegative measurable function on $X_{\ell+1} \times \ldots \times X_N$, which does not vanish $\mu_{\ell+1} \otimes \ldots \otimes \mu_N$-almost everywhere, and let $\overline{M}_\ell$ be the associated multiplication operator. Then the linear operator
$$L = \sum_{\ell=1}^{N} I^{\otimes \ell - 1} \otimes L_\ell \otimes \overline{M}_\ell$$
admits a spectral gap.

EXAMPLE A.4. *Let μ and ν be two probability measures on $\mathbb{R}$, each satisfying a Poincaré inequality. Equip $\mathbb{R}^2$ with the tensor measure $\mu \otimes \nu(dx\,dy) = \mu(dx)\,\nu(dy)$. Then*
$$L = -(\partial_x^* \partial_x + x^2 \partial_y^* \partial_y)$$
is coercive on $L^2(\mu \otimes \nu)/\mathbb{R}$.

A.19. SOME CRITERIA FOR POINCARÉ INEQUALITIES

PROOF OF THEOREM A.2. Let P_ℓ be the orthogonal projection on $(\operatorname{Ker} L_\ell)^\perp$ in $\mathcal{H}_\ell$. The spectral gap assumption means $L_\ell \geq \kappa_\ell P_\ell$. Let M be the multiplication operator by m, then $\overline{M} \geq M$.

When applied to nonnegative operators, tensorization preserves the order: when $A \geq A' \geq 0$ and $B \geq B' \geq 0$, one has $A \otimes B \geq A \otimes B' \geq A' \otimes B'$. Thus,
$$L_1 \otimes \overline{M} + I \otimes L_2 \geq \kappa_1 P_1 \otimes M + \kappa_2 I \otimes P_2.$$
So it is sufficient to prove the theorem when $L_\ell = P_\ell$ and $\overline{m} = m \in L^2(\mu_2)$.

Let $(e_i^1)_{i \geq 0}$ be an orthonormal basis for $\mathcal{H}_1$, such that $(e_i^1)_{i \leq k_1}$ is an orthonormal basis of $\mathcal{K}_1 := \operatorname{Ker} L_1$; and let $(e_j^2)_{j \geq 0}$ be an orthonormal basis for $\mathcal{H}_2$, such that $(e_j^2)_{j \leq k_2}$ is an orthonormal basis of $\mathcal{K}_2 := \operatorname{Ker} L_2$. Then $(e_i^1 \otimes e_j^2)_{i,j \geq 0}$ is an orthonormal basis for $\mathcal{H}$. Moreover, the kernel of L is the vector space generated by $(e_i^1 \otimes e_j^2)_{i,j \in \mathcal{K}}$, where $\mathcal{K} := \{(i,j); \ i \leq k_1, j \leq k_2\}$. So the goal is to prove
$$f = \sum_{(i,j)} c_{ij} e_i^1 \otimes e_j^2 \Longrightarrow$$
$$\kappa_1 \langle (P_1 \otimes M)f, f \rangle + \kappa_2 \langle (I \otimes P_2)f, f \rangle \geq \kappa \sum_{(i,j) \notin \mathcal{K}} c_{ij}^2.$$

First of all,
$$(I \otimes P_2)f = \sum_{(i,j)} c_{ij} e_i^1 \otimes P_2 e_j^2 = \sum_{i \geq 0, \ j \geq k_2+1} c_{ij} e_i^1 \otimes P_2 e_j^2;$$
so
$$(A.19.7) \qquad \langle (I \otimes P_2)f, f \rangle = \sum_{i \geq 0; \ j \geq k_2+1} c_{ij}^2.$$

Next,
$$\langle (P_1 \otimes M)f, f \rangle = \sum_{(i,i',j,j')} c_{ij} c_{i'j'} \langle P_1 e_i^1, e_{i'}^1 \rangle \langle M e_j^2, e_{j'}^2 \rangle$$
$$= \sum_{i \geq k_1+1; \ j,j' \geq 0} c_{ij} c_{ij'} \langle M e_j^2, e_{j'}^2 \rangle$$
$$= \sum_{i \geq k_1+1; \ j,j' \geq k_2+1} c_{ij} c_{ij'} \langle M e_j^2, e_{j'}^2 \rangle$$
$$\quad + 2 \sum_{i \geq k_1+1; \ j \geq k_2+1; \ j' \leq k_2} c_{ij} c_{ij'} \langle M e_j^2, e_{j'}^2 \rangle$$
$$\quad + \sum_{i \geq k_1+1; \ j,j' \leq k_2} c_{ij} c_{ij'} \langle M e_j^2, e_{j'}^2 \rangle.$$

We shall estimate these three sums one after the other:

- The first sum might be rewritten as
$$\sum_{i \geq k_1+1} \left\langle M \left(\sum_{j \geq k_2+1} c_{ij} e_j^2 \right), \left(\sum_{j \geq k_2+1} c_{ij} e_j^2 \right) \right\rangle,$$
and is therefore nonnegative.

- Similarly, the third sum might be rewritten as
$$\sum_{i \geq k_1+1} \left\langle M \left(\sum_{j \leq k_2} c_{ij} e_j^2 \right), \left(\sum_{j \leq k_2} c_{ij} e_j^2 \right) \right\rangle,$$

which can be bounded below by

$$\lambda \sum_{i \geq k_1+1; \, j \leq k_2} c_{ij}^2.$$

- Finally, by applying the inequality $\|Me_j\| \leq \Lambda$ ($j \leq k_2$) and the Cauchy–Schwarz inequality twice, one can bound the second sum from below by

$$-2 \sum_{j \leq k_2} \sum_{i \geq k_1+1} c_{ij} \|Me_j^2\| \left\| \sum_{j' \geq k_2+1} c_{ij'} e_{j'}^2 \right\|$$

$$\geq -2\Lambda \sum_{j \leq k_2} \sqrt{\sum_{i \geq k_1+1} c_{ij}^2} \sqrt{\sum_{i \geq k_1+1} \left\| \sum_{j' \geq k_2+1} c_{ij'} e_{j'}^2 \right\|^2}$$

$$\geq -2\Lambda \sqrt{\sum_{i \geq k_1+1; \, j \leq k_2} c_{ij}^2} \sqrt{\sum_{i \geq k_1+1; \, j \geq k_2+1} c_{ij}^2}.$$

All in all,

$$\langle (P_1 \otimes M)f, f \rangle$$

$$\geq \lambda \sum_{i \geq k_1+1; \, j \leq k_2} c_{ij}^2 - 2\Lambda \sqrt{\sum_{i \geq k_1+1; \, j \leq k_2} c_{ij}^2} \sqrt{\sum_{i \geq k_1+1; \, j \geq k_2+1} c_{ij}^2}$$

$$\geq \frac{\lambda}{2} \sum_{i \geq k_1+1; \, j \leq k_2} c_{ij}^2 - \frac{4\Lambda^2}{\lambda} \sum_{i \geq k_1+1; \, j \geq k_2+1} c_{ij}^2.$$

Combining this with (A.19.7), we see that for all $\theta \in [0,1]$,

$$\langle Lf, f \rangle \geq \kappa_2 \langle (I \otimes P_2)f, f \rangle + \kappa_1 \theta \langle (P_1 \otimes M)f, f \rangle$$

$$\geq \kappa_2 \sum_{i \geq 0; \, j \geq k_2+1} c_{ij}^2 + \frac{\kappa_1 \theta \lambda}{2} \sum_{i \geq k_1+1; \, j \leq k_2} c_{ij}^2 - \frac{4\kappa_1 \theta \Lambda^2}{\lambda} \sum_{i \geq k_1+1; \, j \geq k_2+1} c_{ij}^2$$

$$\geq \kappa \left(\sum_{i \geq 0; \, j \geq k_2+1} c_{ij}^2 + \sum_{i \geq k_1+1; \, j \leq k_2} c_{ij}^2 \right),$$

with

$$\kappa := \min\left(\kappa_2 - \frac{4\kappa_1 \theta \Lambda^2}{\lambda}, \, \frac{\kappa_1 \theta \lambda}{2} \right).$$

To conclude the proof of Theorem A.2, it suffices to choose

$$\theta := \min\left(1, \, \frac{\kappa_2 \lambda}{8 \kappa_1 \Lambda^2} \right). \qquad \square$$

PROOF OF THEOREM A.3. Let M be the multiplication operator by m. The restriction of M to constant functions is obviously coercive with constant $\lambda := \int m \, d\mu_2$, and M is bounded by $\|m\|_{L^1} I$. Then (i) follows by a direct application of Theorem A.2. After that, statement (ii) follows from (i) by a simple induction on N. $\qquad \square$

A.20. Well-posedness for the Fokker–Planck equation

The goal of this Appendix is the following uniqueness theorem:

THEOREM A.5. *With the notation of Theorem 6, for any $f_0 \in L^2((1+E)\,dv\,dx)$, the Fokker–Planck equation (2.7) admits at most one distributional solution $f = f(t,x,v) \in C(\mathbb{R}_+; \mathcal{D}'(\mathbb{R}_x^n \times \mathbb{R}_v^n)) \cap L^\infty_{\mathrm{loc}}(\mathbb{R}_+; L^2((1+E)\,dv\,dx)) \cap L^2_{\mathrm{loc}}(\mathbb{R}_+; H^1_v(\mathbb{R}_x^n \times \mathbb{R}_v^n))$, such that $f(0,\cdot) = f_0$.*

REMARK A.6. The a priori estimates
$$\frac{d}{dt}\int f^2\,dv\,dx = -2\int |\nabla_v f|^2\,dv\,dx + n\int f^2\,dv\,dx$$
$$\frac{d}{dt}\int f^2 E\,dv\,dx = -2\int |\nabla_v f|^2\,dv\,dx + n\int f^2(1+E)\,dv\,dx$$

allow to prove *existence* of a solution, too, for an initial datum $f_0 \in L^2((1+E)\,dv\,dx)$; but this is not what we are after here. (Actually, an existence theorem can be established under much more general assumptions.)

Before going on with the argument, I should explain why the uniqueness statement in Theorem 7 implies the one in Theorem 6. In that case, Proposition 5(iii) can be applied even if ∇V is only continuous: indeed, the differential operator $\nabla V(x) \cdot \nabla_v$ always makes distributional sense. So, if h is any solution of (2.6), satisfying the assumptions of Theorem 6, then $f := h\rho_\infty$ defines a solution of (2.7), and it also satisfies the assumptions in Theorem 7, in view of the inequalities

$$\int f^2(1+E)\,dv\,dx \leq \int f^2 e^E\,dv\,dx = \int h^2 e^{-E}\,dv\,dx,$$

$$\int |\nabla_v f|^2\,dv\,dx = \int |\nabla_v(\rho_\infty h)|^2\,dv\,dx$$
$$\leq 2\int |\nabla_v h|^2 \rho_\infty^2\,dv\,dx + 2\int h^2|\nabla_v \rho_\infty|^2\,dv\,dx$$
$$\leq C \sup_{x,v}\left[(1+|v|^2)e^{-V(x)}e^{-\frac{|v|^2}{2}}\right]\left(\int |\nabla_v h|^2\,d\mu + \int h^2\,d\mu\right).$$

PROOF OF THEOREM A.5. By linearity, it is enough to prove
$$\|f(T,\cdot)\|_{L^2} \leq e^{CT}\|f(0,\cdot)\|_{L^2},$$

which will also yield short-time stability. So let f solve the Fokker–Planck equation in distribution sense, and let $T > 0$ be an arbitrary time.

For any $\varphi \in C^\infty(t,x,v)$, compactly supported in $(0,T) \times \mathbb{R}_x^n \times \mathbb{R}_v^n$, one has
$$\int\int f\left(\partial_t \varphi + v \cdot \nabla_x \varphi - \nabla V(x) \cdot \nabla_v \varphi + \Delta_v \varphi - v \cdot \nabla_v \varphi\right)dt\,dv\,dx = 0.$$

Since $f \in L^\infty([0,T]; L^2(\mathbb{R}_x^n \times \mathbb{R}_v^n))$, a standard approximation procedure shows that

(A.20.1) $\quad \int f(T,\cdot)\varphi(T,\cdot)\,dv\,dx - \int f(0,\cdot)\varphi(0,\cdot)\,dv\,dx =$
$$\int\int f\left(\partial_t \varphi + v \cdot \nabla_x \varphi - \nabla V(x) \cdot \nabla_v \varphi + \Delta_v \varphi - v \cdot \nabla_v \varphi\right)dt\,dv\,dx$$

for all $\varphi \in C^1((0,T); C_c^2(\mathbb{R}_x^n \times \mathbb{R}_v^n)) \cap C([0,T]; C_c^2(\mathbb{R}_x^n \times \mathbb{R}_v^n))$, where $\int\int$ stands for the integral over $[0,T] \times \mathbb{R}_x^n \times \mathbb{R}_v^n$.

Let χ, η be C^∞ functions on $\mathbb{R}^n$ with $0 \le \chi \le 1$, $\chi(x) \equiv 1$ for $|x| \le 1$, $\chi(x) \equiv 0$ for $|x| \ge 2$, $\eta \ge 0$, $\int \eta = 1$, η radially symmetric, $\eta(x) \equiv 0$ for $|x| \ge 1$. With the notation $\varepsilon = (\varepsilon_1, \varepsilon_2)$, $\delta = (\delta_1, \delta_2)$. Define

$$\chi_\varepsilon(x, v) = \chi(\varepsilon_1 x)\,\chi(\varepsilon_2 v), \qquad \eta_\delta(x, v) = \eta(\delta_1 x)\,\eta(\delta_2 v).$$

In words: χ_ε is a family of smooth cut-off functions, and η_δ is a family of mollifiers. (The introduction of η_δ is the main modification with respect to the argument in [**32**, Proposition 5.5].)

Define now

$$f_{\varepsilon,\delta} := (f\chi_\varepsilon) * \eta_\delta, \qquad \varphi_{\varepsilon,\delta} := \chi_\varepsilon((f\chi_\varepsilon) * \eta_\delta * \eta_\delta).$$

The goal is of course to let $\delta \to 0$, $\varepsilon \to 0$ in a suitable way.

Since η is radially symmetric, the identity $\int g(f * \eta) = \int (g * \eta) f$ holds true. So, for any $t \in [0, T]$,

$$(A.20.2) \qquad \int f(t, \cdot)\varphi_{\varepsilon,\delta}(t, \cdot)\, dv\, dx = \int f_{\varepsilon,\delta}(t, \cdot)^2\, dv\, dx.$$

Similarly,

$$(A.20.3) \qquad \int f \partial_t \varphi_{\varepsilon,\delta}\, dv\, dx = \int f_{\varepsilon,\delta}\, \partial_t f_{\varepsilon,\delta}\, dv\, dx = \frac{1}{2}\frac{d}{dt} \int f_{\varepsilon,\delta}^2\, dv\, dx.$$

By combining (A.20.2) and (A.20.3), we get

$$\iint f \partial_t \varphi_{\varepsilon,\delta}\, dv\, dx\, dt = \frac{1}{2}\left(\int f(T, \cdot)\varphi_{\varepsilon,\delta}(T, \cdot)\, dv\, dx - \int f(0, \cdot)\varphi_{\varepsilon,\delta}(0, \cdot)\, dv\, dx \right).$$

So, by plugging $\varphi = \varphi_{\varepsilon,\delta}$ into (A.20.1), one obtains

$$(A.20.4) \qquad \frac{1}{2}\left(\int f(T, \cdot)\varphi(T, \cdot)\, dv\, dx - \int f(0, \cdot)\varphi(0, \cdot)\, dv\, dx \right)$$

$$(A.20.5) \qquad = \iint f\chi_\varepsilon(v \cdot \nabla_x - \nabla V(x) \cdot \nabla_v + \Delta_v - v \cdot \nabla_v)(f_{\varepsilon,\delta} * \eta_\delta)\, dv\, dx\, dt$$

$$(A.20.6) \qquad + \iint f\Big[(v \cdot \nabla_x - \nabla V(x) \cdot \nabla_v + \Delta_v - v \cdot \nabla_v)\chi_\varepsilon\Big](f_{\varepsilon,\delta} * \eta_\delta)\, dv\, dx\, dt$$

$$(A.20.7) \qquad + 2\iint f\chi_\varepsilon \nabla_v \chi_\varepsilon \cdot \nabla_v (f_{\varepsilon,\delta} * \eta_\delta)\, dv\, dx\, dt.$$

For any given $\varepsilon > 0$, all the functions involved are restricted to a compact set K_ε in the variable $X = (x, v)$, uniformly in $\delta \le 1$. Now use the identities $\nabla(g * \eta) = (\nabla g) * \eta$, $\Delta(g * \eta) = (\Delta g) * \eta$, $\int g(h * \eta) = \int h(g * \eta)$ to rewrite (A.20.5) as

$$(A.20.8) \qquad \iint f_{\varepsilon,\delta}(v \cdot \nabla_x - \nabla V(x) \cdot \nabla_v + \Delta_v - v \cdot \nabla_v) f_{\varepsilon,\delta}\, dv\, dx\, dt$$

$$+ \iint f\chi_\varepsilon \Big[\xi \cdot \nabla(f_{\varepsilon,\delta} * \eta_\delta) - (\xi \cdot \nabla f_{\varepsilon,\delta}) * \eta_\delta \Big],$$

where ξ is a temporary notation for the vector field $(v, -\nabla V(x) - v)$. By integration by parts, the first integral in (A.20.8) can be rewritten as

$$(A.20.9) \quad -\iint |\nabla_v f_{\varepsilon,\delta}|^2 - \frac{1}{2} \iint (v \cdot \nabla_v) f_{\varepsilon,\delta}^2 = -\iint |\nabla_v f_{\varepsilon,\delta}|^2 + \frac{n}{2} \iint f_{\varepsilon,\delta}^2.$$

Now we should bound

$$(A.20.10) \quad \left\| \xi \cdot \nabla (f_{\varepsilon,\delta} * \eta_\delta) - (\xi \cdot \nabla f_{\varepsilon,\delta}) * \eta_\delta \right\|_{L^2}$$
$$= \left\| \int [\xi(X) - \xi(Y)] \cdot \nabla f_{\varepsilon,\delta}(Y) \, \eta_\delta(Y - X) \right\|_{L^2(dX)}.$$

We shall estimate the contributions of $v \cdot \nabla_x$, $v \cdot \nabla_v$ and $\nabla_x V \cdot \nabla_v$ separately. First, with obvious notation,

$$\left\| [v \cdot \nabla_x, \eta_\delta *] f_{\varepsilon,\delta} \right\|_{L^2}$$
$$= \left\| \int (v - w) \cdot \nabla_x f_{\varepsilon,\delta}(y, w) \, \eta_{\delta_1}(x - y) \, \eta_{\delta_2}(v - w) \, dw \, dy \right\|_{L^2}$$
$$= \left\| \int f_{\varepsilon,\delta}(y, w) \, (v - w) \cdot \nabla_x \eta_{\delta_1}(x - y) \, \eta_{\delta_2}(v - w) \, dw \, dy \right\|_{L^2}.$$

Inside the integral, one has $|v - w| \leq \delta_2$, $|x - y| \leq \delta_1$, and also $|\nabla_x \eta_{\delta_1}| = O(\delta_1^{-(n+1)})$, $\eta_{\delta_2} = O(\delta_2^{-n})$; so, all in all,

$$\left\| [v \cdot \nabla_x, \eta_\delta *] f_{\varepsilon,\delta} \right\|_{L^2} \leq C \frac{\delta_2}{\delta_1} \left\| \frac{1}{\delta_1^n \delta_2^n} \int f_{\varepsilon,\delta}(y, w) \, 1_{|x-y| \leq \delta_1} 1_{|v-w| \leq \delta_2} \, dw \, dy \right\|_{L^2}$$
$$\leq C \frac{\delta_2}{\delta_1} \| f_{\varepsilon,\delta} \|_{L^2}$$
$$\leq C \frac{\delta_2}{\delta_1} \| f \|_{L^2},$$

where the last two inequalities follow from Young's convolution inequality.

Next,

$$\left\| [v \cdot \nabla_v, \eta_\delta *] f_{\varepsilon,\delta} \right\|_{L^2}$$
$$= \left\| \int (v - w) \cdot \nabla_v f_{\varepsilon,\delta}(y, w) \, \eta_{\delta_1}(x - y) \, \eta_{\delta_2}(v - w) \, dw \, dy \right\|.$$

Using the fact that $|v - w| \leq \delta_2$ inside the integral and applying Young's convolution inequality as before, we find

$$\left\| [v \cdot \nabla_v, \eta_\delta *] f_{\varepsilon,\delta} \right\|_{L^2} \leq C \delta_2 \| \nabla_v f_{\varepsilon,\delta} \|_{L^2} \leq C \delta_2 \| \nabla_v f \|_{L^2}.$$

Finally,

$$\left\| [\nabla_x V \cdot \nabla_v, \eta_\delta *] f_{\varepsilon,\delta} \right\|_{L^2}$$
$$= \left\| \int [\nabla V(x) - \nabla V(y)] \cdot \nabla_v f_{\varepsilon,\delta}(y, w) \, \eta_{\delta_1}(x - y) \, \eta_{\delta_2}(v - w) \, dw \, dy \right\|_{L^2}$$
$$\leq C \, \sup \left\{ |\nabla V(x) - \nabla V(y)|; \, |x - y| \leq \delta_1; \, x, y \in K_{\varepsilon_1} \right\} \| \nabla_v f_{\varepsilon,\delta} \|_{L^2}$$
$$\leq C \, \theta_{\varepsilon_1}(\delta_1) \| \nabla_v f_{\varepsilon,\delta} \|_{L^2},$$

where θ_ε stands for the modulus of continuity of ξ on the compact set K_ε. In all these estimates, the L^2 norm was taken with respect to all variables t, x, v. The conclusion is that the L^2 norm in (A.20.10) is bounded like

$$\text{(A.20.11)} \qquad O\left(\frac{\delta_2}{\delta_1}\|f\|_{L^2}^2 + \delta_2\|\nabla_v f\|_{L^2} + \theta_{\varepsilon_1}(\delta_1)\|\nabla_v f\|_{L^2}\right).$$

Next, since $\|\nabla_v \chi_\varepsilon\|_{L^\infty} \le C\varepsilon_2$, it is possible to bound (A.20.7) by

$$\text{(A.20.12)} \quad C\varepsilon_2 \|f\chi_\varepsilon\|_{L^2} \|\nabla_v(f_{\varepsilon,\delta} * \eta_\delta)\|_{L^2} \le C\varepsilon_2 \|f\|_{L^2}(\|\nabla_v f\|_{L^2} + \|f\|_{L^2}).$$

Finally, the terms in the integrand of (A.20.6) can be bounded with the help of the inequalities

$$|v \cdot \nabla_x \chi_\varepsilon(x,v)| \le C|v|\varepsilon_1, \qquad |\nabla V(x) \cdot \nabla_v \chi_\varepsilon(x,v)| \le C\varepsilon_2 M(\varepsilon_1^{-1}),$$
$$|\Delta_v \chi_\varepsilon(x,v)| \le C\varepsilon_2^2, \qquad |v \cdot \nabla_v \chi_\varepsilon(x,v)| \le C|v|\varepsilon_2,$$

where $M(R) := \sup\{|\nabla V(x)|; |x| \le 2R\}$. Then, by Cauchy–Schwarz again, (A.20.6) can be bounded by

$$C\Big[\varepsilon_1 + \varepsilon_2(1 + M(\varepsilon_1^{-1}))\Big] \sqrt{\iint f^2 |v|^2 \, dv \, dx \, dt} \sqrt{\iint (f_{\varepsilon,\delta} * \eta_\delta)^2 \, dv \, dx \, dt}.$$

Since $|v|^2 \le 2E - 2(\inf V)$, in the end (A.20.6) is controlled by

$$\text{(A.20.13)} \qquad C\Big[\varepsilon_1 + \varepsilon_2(1 + M(\varepsilon_1^{-1}))\Big] \sqrt{\iint f^2 E \, dv \, dx \, dt} \sqrt{\iint f^2 \, dv \, dx \, dt}.$$

Plugging the bounds (A.20.9), (A.20.13) and (A.20.12) into (A.20.4), we conclude that

$$\text{(A.20.14)} \quad \frac{1}{2}\left(\int f_{\varepsilon,\delta}^2(T,x,v) \, dv \, dx - \int f_{\varepsilon,\delta}^2(0,x,v) \, dv \, dx\right)$$
$$\le -\iint |\nabla_v f_{\varepsilon,\delta}|^2(t,x,v) \, dv \, dx \, dt + \frac{n}{2}\iint f_{\varepsilon,\delta}^2(t,x,v) \, dv \, dx \, dt$$
$$+ C\left(\frac{\delta_2}{\delta_1}\|f\|_{L^2}^2 + \delta_2\|f\|_{L^2}\|\nabla_v f_{\varepsilon,\delta}\|_{L^2} + \theta_{\varepsilon_1}(\delta_1)\|\nabla_v f_{\varepsilon,\delta}\|_{L^2}\|f_{\varepsilon,\delta}\|_{L^2}\right)$$
$$+ C(\varepsilon_1 + \varepsilon_2 M(\varepsilon_1^{-1}))\|f(1+E)\|_{L^2}^2,$$

where all the L^2 norms in the right-hand side are with respect to $dv \, dx \, dt$. Now let $\delta_2 \to 0$, then $\delta_1 \to 0$, then $\varepsilon_2 \to 0$, then $\varepsilon_1 \to 0$, then $\delta \to 0$: all the error terms in the right-hand side of (A.20.14) vanish in the limit, and $f_{\varepsilon,\delta}$ converges to f almost everywhere and in $L^2(dv \, dx \, dt)$. So

$$\int f^2(T,x,v) \, dv \, dx \le \liminf \int f_{\varepsilon,\delta}^2(T,x,v) \, dv \, dx$$
$$\le \liminf \left[\int f_{\varepsilon,\delta}^2(0,x,v) \, dv \, dx + \frac{n}{2}\iint f_{\varepsilon,\delta}^2(t,x,v) \, dv \, dx \, dt\right]$$
$$= \int f^2(0,x,v) \, dv \, dx + \frac{n}{2}\iint f^2(t,x,v) \, dv \, dx \, dt.$$

By Gronwall's lemma,

$$\|f(t,\cdot)\|_{L^2(\mathbb{R}^n \times \mathbb{R}^n)} \le e^{\frac{nt}{4}} \|f(0,\cdot)\|_{L^2(\mathbb{R}^n \times \mathbb{R}^n)},$$

which concludes the argument. □

REMARK A.7. Just as in [**32**, Proposition 5.5], the particular structure of the Fokker–Planck equation was mainly used in the estimate $|\nabla V(x) \cdot \nabla_v \chi_\varepsilon| \leq \varepsilon_2 M(\varepsilon_1^{-1})$. It would be interesting to understand to what extent this computation can be generalized to larger classes of linear equations, and whether this has anything to do with the hypoelliptic structure.

A.21. Some methods for global hypoellipticity

This Appendix is devoted to various hypoelliptic regularization estimates for the Fokker–Planck equation. To get a good intuition of the equation, or to test the optimality of certain results, it is good to remember that the fundamental solution can be explicitly computed when the confining potential is quadratic; see e.g. [**47**, pp. 238–240] or [**14**, Section 5]. For the convenience of the reader I shall recall the result in the simple case when there is neither confining nor friction (this simplification does not alter the short-time properties of the equation): it was worked out by Kolmogorov [**39**] that the fundamental solution of the operator $\partial_t + v \cdot \nabla_x f - \theta \Delta_v$, starting from the initial measure $\delta_{(x_0, v_0)}$, is

$$(3\pi^2 \theta)^{-n} t^{-2n} \exp\left[-\frac{1}{\pi^2 \theta} \left(\frac{3|x - (x_0 + tv_0)|^2}{t^3} - \frac{3[x - (x_0 + tv_0)] \cdot (v - v_0)}{t^2} + \frac{|v - v_0|^2}{t} \right)\right].$$

For heuristic purposes this solution may be replaced by

$$(A.21.1) \qquad t^{-2n} \exp\left[-c \left(\frac{|x - (x_0 + tv_0)|^2}{t^3} + \frac{|v - v_0|^2}{t} \right)\right].$$

From this representation one can get many short-time smoothness estimates, but of course this will work only for quadratic potentials, or perturbations thereof [**14**, Section 5]. Then the problem arises to recover as much as possible of these estimates without using the fundamental solution, but using instead robust functional tools.

Previous work in this area has been done by a number of authors, but they were mainly concerned with local estimates [**36, 38, 49**]; the most notable exceptions are the recent paper by Bouchut [**7**] and the works by Helffer, Hérau and Nier [**32, 34**]. The history of the subject is reviewed in those references. The methods considered in the present Appendix seem to be new, and can certainly be extended to more general classes of equations; they are mainly based on a combination of differential inequalities and functional inequalities. At the same time as I was working on this problem, Hérau [**33**] was devising related methods, which I will discuss later.

I shall consider three particular estimates (those which were used in the present paper): First, the $L^2(\mu) \to H^1(\mu)$ regularization for the Fokker–Planck equation in the form (7.1); secondly, the $M \to H^k$ regularization for the Fokker–Planck equation in the form (7.9) (here M is the space of bounded measures, and H^k is the *non-weighted* Sobolev space of order k); and thirdly, an entropic regularization effect of the form $L \log L \to L|\nabla \log L|^2$.

A.21.1. From weighted L^2 to weighted H^1.

In this subection, V is a C^2 potential on $\mathbb{R}^n$, bounded below, $\gamma(v) = (2\pi)^{-n/2} e^{-|v|^2/2}$ is the standard Gaussian, and $\mu(dx\,dv) = \gamma(v) e^{-V(x)}\,dv\,dx$ stands for the equilibrium measure associated with the Fokker–Planck equation (7.1) (it might have finite or infinite mass). Apart from that, the only regularity assumption is the existence of a constant C such that

$$(A.21.2) \qquad |\nabla^2 V| \leq C(1 + |\nabla V|).$$

As we shall see, this is sufficient to get estimate (7.8), independently of the fact that e^{-V} satisfies the Poincaré inequality (7.5) or not.

THEOREM A.8. *Let V be a C^2 function on $\mathbb{R}^n$, bounded below and satisfying* (A.21.2). *Then, solutions of the Fokker–Planck equation* (7.1) *with initial datum h_0 satisfy*

$$0 \leq t \leq 1 \implies \|\nabla_x h(t,\cdot)\|_{L^2(\mu)} + \sum_{k=1}^{3} \|\nabla_v^k h(t,\cdot)\|_{L^2(\mu)} \leq \frac{C}{t^{3/2}} \|h_0\|_{L^2(\mu)}$$

for some constant C, only depending on n and on the constant C appearing in (A.21.2).

REMARK A.9. These estimates also seem to be new. The proof can be adapted to cover the case of L^1 initial data, at the price of a deterioration of the exponents. I shall explain this later on.

REMARK A.10. Theorem A.8 shows that (with obvious notation) e^{-tL} maps L^2 into $H_x^1 \cap H_v^3$ with norm $O(t^{-3/2})$. It also maps L^2 into L^2 with norm $O(1)$; so, by interpolation, it maps L^2 into $H_x^\alpha \cap H_v^{3\alpha}$ with norm $O(t^{-3\alpha/2})$, for all $\alpha \in [0,1]$. For $\alpha = 2/3$ this shows

$$\left\| D\, t\, e^{-t(I+L)} \right\|_{L^2 \to L^2} \leq C, \qquad D = (-\Delta_v)^{1/2} + (-\Delta_x)^{1/6}.$$

Since $\int_0^1 e^{-t(1+L)}\,dt$ is a parametrix for $(I+L)^{-1}$, and $t^{-\beta}$ is integrable at $t=0$ for $\beta < 1$, one can deduce a "stationary" hypoelliptic regularity estimate à la Kohn:

$$(A.21.3) \qquad \|h\|_{H_x^\alpha(\mu)} + \|h\|_{H_v^{3\alpha}(\mu)} \leq C\bigl(\|h\|_{L^2(\mu)} + \|Lh\|_{L^2(\mu)}\bigr), \qquad \forall \alpha < 2/3.$$

With a more refined analysis, it may be possible to catch the optimal exponent $\alpha = 2/3$ in the above estimate, but I shall not explore this somewhat tricky issue here.

PROOF OF THEOREM A.8. As a consequence of Theorem 6 and an approximation argument which is omitted here, it is sufficient to prove this theorem for smooth, rapidly decaying solutions. So I shall not worry about technical justification of the manipulations below. Also, C will stand for various constants which only depend on n and the constant in (A.21.2).

The following estimates will be used several times. As a consequence of Lemma A.24 in Appendix A.23, for each v,

$$\int_{\mathbb{R}^n} |\nabla V(x)|^2 g^2(x,v) e^{-V(x)}\,dx$$

$$\leq C \left(\int g^2(x,v) e^{-V(x)}\,dx + \int |\nabla_x g(x,v)|^2 e^{-V(x)}\,dx \right);$$

by integrating this with respect to $\gamma(v)\,dv$ one obtains

(A.21.4) $$\int_{\mathbb{R}^n \times \mathbb{R}^n} |\nabla V|^2 g^2 \, d\mu \leq C \left(\int g^2 \, d\mu + \int |\nabla_x g|^2 \, d\mu \right).$$

Similarly,

(A.21.5) $$\int_{\mathbb{R}^n \times \mathbb{R}^n} |v|^2 g^2 \, d\mu \leq C \left(\int g^2 \, d\mu + \int |\nabla_v g|^2 \, d\mu \right).$$

Now we turn to the main part of the argument, which can be decomposed into four steps.

Step 1: *"Energy" estimate in H_x^1 and H_v^3 norms combined.*

To avoid heavy notation, I shall use symbolic matrix notation which should be rather self-explanatory, and write

$$L = v \cdot \nabla_x - \nabla V(x) \cdot \nabla_v - \Delta_v - v \cdot \nabla_v.$$

By differentiating the equation once with respect to x, and three times with respect to v, one finds

(A.21.6) $$\left(\frac{\partial}{\partial t} + L \right) \nabla_x h = \nabla_x^2 V(x) \cdot \nabla_v h,$$

(A.21.7) $$\left(\frac{\partial}{\partial t} + L \right) \nabla_v^3 h = -3 \nabla_v^2 \nabla_x h - 3 \nabla_v^3 h.$$

After taking the scalar product of (A.21.6) by $\nabla_x h$ and integrating against μ, we get

(A.21.8) $$\frac{1}{2} \frac{d}{dt} \int |\nabla_x h|^2 \, d\mu + \int |\nabla_v \nabla_x h|^2 \, d\mu = \int (\nabla_x^2 V) \nabla_v h \cdot \nabla_x h \, d\mu.$$

Similarly, from (A.21.7) it follows that

(A.21.9) $$\frac{1}{2} \frac{d}{dt} \int |\nabla_v^3 h|^2 \, d\mu + \int |\nabla_v^4 h|^2 \, d\mu = -3 \int |\nabla_v^3 h|^2 \, d\mu - 3 \int \nabla_v^3 h \cdot \nabla_v^2 \nabla_x h \, d\mu.$$

Let us bound the right-hand side of (A.21.8). Since $(\nabla_v)^* = -\nabla_v + v$ (where the $*$ is for the adjoint in $L^2(\mu)$), one has

$$\int (\nabla_x^2 V) \nabla_v h \cdot \nabla_x h \, d\mu = -\int (\nabla_x^2 V) h \cdot \nabla_x \nabla_v h \, d\mu - \int \langle (\nabla_x^2 V) h v, \nabla_x h \rangle \, d\mu.$$

By Cauchy–Schwarz and Young's inequality,

$$-\int (\nabla_x^2 V) h \cdot \nabla_x \nabla_v h \, d\mu \leq \int |\nabla_x^2 V|^2 h^2 \, d\mu + \frac{1}{4} \int |\nabla_x \nabla_v h|^2 \, d\mu.$$

Thanks to (A.21.4), this can be bounded by

$$C \left(\int |\nabla_x h|^2 \, d\mu + \int h^2 \, d\mu \right) + \frac{1}{4} \int |\nabla_x \nabla_v h|^2 \, d\mu.$$

By Cauchy–Schwarz inequality again,

(A.21.10) $$-\int \langle (\nabla_x^2 V) h v, \nabla_x h \rangle \, d\mu \leq \sqrt{\int |\nabla_x^2 V|^2 h^2 \, d\mu} \sqrt{\int |v|^2 |\nabla_x h|^2 \, d\mu}.$$

In view of (A.21.5),

(A.21.11) $$\frac{1}{2} \int |v|^2 |\nabla_x h|^2 \, d\mu \leq C \left(\int |\nabla_x h|^2 \, d\mu + \int |\nabla_v \nabla_x h|^2 \, d\mu \right).$$

By (A.21.10), (A.21.11) and Young's inequality, there is a constant C such that

$$-\int \langle (\nabla_x^2 V)hv, \nabla_x h \rangle \, d\mu \leq C \left(\int |\nabla_x^2 V|^2 h^2 \, d\mu + \int |\nabla_x h|^2 \, d\mu \right)$$
$$+ \frac{1}{4} \int |\nabla_v \nabla_x h|^2 \, d\mu.$$

All in all,

(A.21.12) $\quad \dfrac{1}{2} \dfrac{d}{dt} \int |\nabla_x h|^2 \, d\mu + \dfrac{1}{2} \int |\nabla_v \nabla_x h|^2 \, d\mu$
$$\leq C \left(\int |\nabla_x^2 V|^2 h^2 \, d\mu + \int |\nabla_x h|^2 \, d\mu \right).$$

The right-hand side in (A.21.9) is estimated in a similar way:

$$-3 \int \nabla_v^3 h \cdot \nabla_v^2 \nabla_x h \, d\mu = 3 \int \nabla_v^4 h \cdot \nabla_v \nabla_x h \, d\mu - 3 \int \nabla_v^3 h \cdot v \nabla_v \nabla_x h \, d\mu;$$

then on one hand

$$3 \int \nabla_v^4 h \cdot \nabla_v \nabla_x h \, d\mu \leq \frac{1}{4} \int |\nabla_v^4 h|^2 \, d\mu + 9 \int |\nabla_v \nabla_x h|^2 \, d\mu;$$

on the other hand, again by (A.21.5),

$$-3 \int \nabla_v^3 h \cdot v \nabla_v \nabla_x h \, d\mu \leq 3 \sqrt{\int |v|^2 |\nabla_v^3 h|^2 \, d\mu} \sqrt{\int |\nabla_v \nabla_x h|^2 \, d\mu}$$
$$\leq C \sqrt{\int |\nabla_v^4 h|^2 \, d\mu + \int |\nabla_v^3 h|^2 \, d\mu} \sqrt{\int |\nabla_v \nabla_x h|^2 \, d\mu}$$
$$\leq \frac{1}{4} \int |\nabla_v^4 h|^2 \, d\mu + \frac{1}{4} \int |\nabla_v^3 h|^2 \, d\mu + C \int |\nabla_v \nabla_x h|^2 \, d\mu.$$

So there is a constant C such that

(A.21.13) $\quad \dfrac{1}{2} \dfrac{d}{dt} \int |\nabla_v^3 h|^2 \, d\mu + \dfrac{1}{2} \int |\nabla_v^4 h|^2 \, d\mu$
$$\leq C \left(\int |\nabla_v \nabla_x h|^2 \, d\mu + \int |\nabla_v^3 h|^2 \, d\mu + \int |\nabla_v^4 h|^2 \, d\mu \right).$$

As a consequence of (A.21.12) and (A.21.13) it is possible to find numerical constants $a, K, C > 0$ (only depending on n and C in (A.21.2)) such that

(A.21.14) $\quad \dfrac{d}{dt} \left(\int |\nabla_x h|^2 \, d\mu + a \int |\nabla_v^3 h|^2 \, d\mu \right)$
$$\leq -K \left(\int |\nabla_v^4 h|^2 \, d\mu + \int |\nabla_v \nabla_x h|^2 \, d\mu \right)$$
$$+ C \left(\int h^2 \, d\mu + \int |\nabla_x h|^2 \, d\mu + \int |\nabla_v^3 h|^2 \, d\mu \right).$$

This concludes the first step.

Step 2: *Time-behavior of the mixed derivative.*

In this step I shall focus on the mixed derivative integral $\int \nabla_x h \cdot \nabla_v h \, d\mu$. By differentiating the equation with respect to x and multiplying by $\nabla_v h$, differentiating the equation with respect to v and multiplying by $\nabla_x h$, then using the chain rule and the identity $F\Delta_v G + G\Delta_v F = \Delta_v(FG) - 2\nabla_v F \cdot \nabla_v G$, one easily obtains

$$\left(\frac{\partial}{\partial t} + L\right)(\nabla_x h \cdot \nabla_v h) = \langle \nabla_x^2 V \cdot \nabla_v h, \nabla_v h\rangle$$
$$- 2\nabla_v \nabla_x h \cdot \nabla_v^2 h \, d\mu - |\nabla_x h|^2 - \nabla_x h \cdot \nabla_v h.$$

After integration against μ, this yields

(A.21.15)
$$\frac{1}{2}\frac{d}{dt}\int \nabla_x h \cdot \nabla_v h \, d\mu = \int \langle \nabla_x^2 V \cdot \nabla_v h, \nabla_v h\rangle \, d\mu - 2\int \nabla_v \nabla_x h \cdot \nabla_v^2 h \, d\mu$$
$$- \int |\nabla_x h|^2 \, d\mu - \int \nabla_x h \cdot \nabla_v h \, d\mu.$$

The first term in the right-hand side need some rewriting: Since $(\nabla_v)^* = -\nabla_v + v$,

$$\int \langle \nabla_x^2 V \cdot \nabla_v h, \nabla_v h\rangle \, d\mu = -\int \nabla_x^2 V h \nabla_v^2 h \, d\mu - \int h \langle \nabla_x^2 V v, \nabla_v h\rangle \, d\mu$$
$$\leq \sqrt{\int |\nabla_x^2 V|^2 h^2 \, d\mu} \sqrt{\int |\nabla_v^2 h|^2 \, d\mu} + \sqrt{\int |\nabla_x^2 V|^2 h^2 \, d\mu} \sqrt{\int |v|^2 |\nabla_v h|^2 \, d\mu}.$$

With the help of Young's inequality and (A.21.5) again, this can be bounded by

$$\varepsilon \int |\nabla_x^2 V|^2 h^2 \, d\mu + C_\varepsilon \left(+\int |\nabla_v^2 h|^2 \, d\mu + \int |\nabla_v h|^2 \, d\mu\right).$$

By Lemma A.24, if ε is small enough then this is bounded by

$$\frac{1}{4}\left(\int |\nabla_x h|^2 \, d\mu + \int h^2 \, d\mu\right) + C\left(\int |\nabla_v h|^2 \, d\mu + \int |\nabla_v^2 h|^2 \, d\mu\right).$$

Now for the second term in the right-hand side of (A.21.15), we just write

$$-2\int \nabla_v \nabla_x h \nabla_v^2 h \, d\mu \leq \int |\nabla_v \nabla_x h|^2 \, d\mu + \int |\nabla_v^2 h|^2 \, d\mu.$$

Summarizing all the above computations: There is a numerical constant C, only depending on n and C in (A.21.2), such that

(A.21.16)
$$\frac{d}{dt}\int \nabla_x h \cdot \nabla_v h \, d\mu \leq -\frac{1}{2}\int |\nabla_x h|^2 \, d\mu$$
$$+ C\left(\int h^2 \, d\mu + \int |\nabla_v h|^2 \, d\mu + \int |\nabla_v^2 h|^2 \, d\mu + \int |\nabla_x \nabla_v h|^2 \, d\mu\right).$$

This concludes the second step of the proof.

REMARK A.11. We could also have conducted the computations in the following way:

$$-2\int \nabla_v \nabla_x h \nabla_v^2 h \, d\mu = 2\int \nabla_x h \cdot \nabla_v^3 h \, d\mu - 2\int \nabla_x h \cdot v \nabla_v^2 h \, d\mu.$$

Then on one hand,

$$2\int \nabla_x h \cdot \nabla_v^3 h \, d\mu \leq \frac{1}{4}\int |\nabla_x h|^2 \, d\mu + 4\int |\nabla_v^3 h|^2 \, d\mu;$$

on the other hand, just as before,

$$-2 \int \nabla_x h \cdot v \nabla_v^2 h \, d\mu$$

$$\leq \frac{1}{4} \int |\nabla_x h|^2 \, d\mu + 4 \int |v|^2 |\nabla_v^2 h|^2 \, d\mu$$

$$\leq \frac{1}{4} \int |\nabla_x h|^2 \, d\mu + C \left(\int |\nabla_v^2 h|^2 \, d\mu + \int |\nabla_v^3 h|^2 \, d\mu \right).$$

By doing so, we would have obtained the same result as (A.21.16), except that the integral $\int |\nabla_x \nabla_v h|^2 \, d\mu$ would be replaced by $\int |\nabla_v^3 h|^2 \, d\mu$. Then the rest of the proof would have worked through.

Step 3: *Interpolation inequalities*

If h is a function of v, lying in $L^2(\gamma)$, one can write $h = \sum_k c_k H_k$, where H_k are normalized Hermite polynomials and k are multi-indices in $\mathbb{N}^n$; then

$$\int h^2 \, d\gamma = \sum_k c_k^2, \qquad \int |\nabla_v h|^2 \, d\mu = \sum |k|^2 c_k^2,$$

$$\int |\nabla_v^2 h|^2 \, d\mu = \sum |k|^4 c_k^2, \qquad \text{etc.}$$

(here $|k|^2 = \sum k_\ell^2$, $1 \leq \ell \leq n$ and $|k|^4 = (|k|^2)^2$). Then, by Hölder's inequality (in the k variable), one can prove interpolation inequalities such as

$$\int |\nabla_v h|^2 \, d\gamma \leq C \left(\int h^2 \, d\gamma \right)^{2/3} \left(\int |\nabla_v^3 h|^2 \, d\gamma \right)^{1/3}.$$

Now if $h = h(x, v)$ is a function of both variables x and v, one can apply the previous inequality to $h(x, \cdot)$ for each x, then integrate with respect to $e^{-V}(x) \, dx$, and apply Hölder's inequality in the x variable, to find

$$\int |\nabla_v h|^2 \, d\mu \leq C \left(\int h^2 \, d\mu \right)^{2/3} \left(\int |\nabla_v^3 h|^2 \, d\mu \right)^{1/3}.$$

Similarly,

$$\int |\nabla_v^j h|^2 \, d\mu \leq C \left(\int h^2 \, d\mu \right)^{1-(j/4)} \left(\int |\nabla_v^4 h|^2 \, d\mu \right)^{j/4}, \qquad 1 \leq j \leq 3.$$

Step 4: *Conclusion*

Now we can turn to the proof of estimate (7.8). Without loss of generality, assume $\int h^2 \, d\mu = 1$ at time 0. Then, since this quantity is nonincreasing with time, $\int h^2(t, \cdot) \, d\mu \leq 1$ for all $t \geq 0$. By combining the results of Steps 1, 2 and 3, we discover that the quantities

$$X := \int |\nabla_x h|^2 \, d\mu, \qquad Y_j := \int |\nabla_v^j h|^2 \, d\mu \qquad (0 \leq j \leq 4),$$

$$\mathcal{M} := \int \nabla_x h \cdot \nabla_v h, \qquad W = \int |\nabla_x \nabla_v h|^2,$$

viewed as functions of t, solve the system of differential inequalities

(A.21.17)
$$\begin{cases} \dfrac{d}{dt}(X + aY_3) \leq -K(Y_4 + W) + C(1 + X + Y_3) \\ \dfrac{d}{dt}\mathcal{M} \leq -KX + C(1 + Y_1 + Y_2 + W) \\ |\mathcal{M}| \leq \sqrt{XY_1}; \quad Y_1 \leq CY_2^{1/2} \leq C'Y_3^{1/3} \leq C''Y_4^{1/4}. \end{cases}$$

It is a consequence of Lemma A.26 in Appendix A.23 that solutions of (A.21.17) satisfy
$$0 \leq t \leq 1 \implies X(t) + Y(t) \leq \frac{A}{t^3}$$
for some computable constant A. As a consequence, for $0 \leq t \leq 1$,
$$\int |\nabla_x h|^2 \, d\mu = O(t^{-3}), \quad \int |\nabla_v^3 h|^2 \, d\mu = O(t^{-3}).$$
Then, by interpolation $\int |\nabla_v h|^2 = O(t^{-1})$, $\int |\nabla_v^2 h|^2 = O(t^{-2})$. This concludes the proof of (7.8). □

A.21.2. Variants. Here I studied the regularization effect by means of a *system* of differential inequalities. It is natural to ask whether one can do the same with just one differential inequality. The answer is affirmative: It is possible to use a trick similar to the one in the proof of Theorem 18, that is, add a carefully chosen lower-order term which is derived from the mixed derivative $\int \nabla_x h \cdot \nabla_v h$.

A first possibility is to consider the Lyapunov functional
$$\mathcal{E}(h) = \int h^2 \, d\mu + a \int |\nabla_x h|^2 \, d\mu + 2b \int \nabla_x(D_x^{1/3} h) \cdot \nabla_v(D_x^{1/3} h) \, d\mu$$
$$+ c \int |\nabla_v^3 h|^2 \, d\mu,$$
where $D_x = (-\Delta_x)^{1/2}$. Then by using computations similar to the ones in Subsection A.21.1, plus estimates on the commutator $[D_x^{1/3}, \nabla V]$, one can establish the following a priori estimate along the Fokker–Planck equation: For well-chosen positive constants a, b, c,
$$\frac{d}{dt}\mathcal{E}(h) \leq -K\mathcal{E}(h)^{4/3}, \quad h = e^{-tL} h_0.$$
The desired result follows immediately.

One drawback of this method is the introduction of fractional derivatives. There is a nice variant due to Hérau [33] in which one avoids this by using powers of t:
$$\mathcal{F}(t, h) = \int h^2 \, d\mu + at \int |\nabla_v h|^2 \, d\mu + 2bt^2 \int \nabla_v h \cdot \nabla_x h + ct^3 \int |\nabla_x h|^2 \, d\mu.$$
Then one can estimate the time-derivative of $\mathcal{F}(t, e^{-tL} h_0)$ by means of computations similar to those in Subsection A.21.1, and the inequalities
$$t \left| \int (\nabla_v h \cdot \nabla_x h) \, d\mu \right| \leq C \int |\nabla_v h|^2 \, d\mu + \varepsilon t^2 \int |\nabla_x h|^2 \, d\mu;$$
$$t^2 \left| \int \nabla_x \nabla_v h \cdot \nabla_v^2 h \, d\mu \right| \leq Ct \int |\nabla_v^2 h|^2 \, d\mu + \varepsilon t^3 \int |\nabla_x \nabla_v h|^2 \, d\mu.$$

In the end, if a, b, c are well-chosen, one obtains, with the shorthand $h = e^{-tL}h_0$,

$$\frac{d}{dt}\mathcal{F}(t,h) \leq -K\left(\int |\nabla_v h|^2\, d\mu + t\int |\nabla_v^2 h|^2\, d\mu + t^2 \int |\nabla_x h|^2\, d\mu \right.$$
$$\left. + t^3 \int |\nabla_x \nabla_v h|^2\, d\mu\right).$$

It follows that $\mathcal{F}(t,h)$ is nonincreasing, and therefore

$$\int |\nabla_v h|^2\, d\mu = O(t^{-1}), \qquad \int |\nabla_x h|^2\, d\mu = O(t^{-3}).$$

The conclusion is not so strong as the one we had before, since we only have estimates on the first-order derivative in v. But the exponents are again optimal, and it is possible to adapt the method and recover estimates on higher-order derivatives. Furthermore, estimates on $\int |\nabla_x h|^2$ and $\int |\nabla_v h|^2$ are exactly what is needed for Theorem 37 to apply.

Hérau's method lends itself very well to an abstract treatment. For instance, let us consider an abstract operator $L = A^*A + B$, satisfying Assumptions (i)–(iii) in Theorem 18, then the following decay rates (in general optimal) can be proven, at least formally:

(A.21.18) $$\|Ae^{-tL}h\| = O(t^{-1/2}); \qquad \|Ce^{-tL}h\| = O(t^{-3/2}).$$

To show this, introduce

$$\mathcal{F}(t,h) := \|e^{-tL}h\|^2 + at\|Ae^{-tL}h\|^2 + 2bt^2\langle Ae^{-tL}h, Ce^{-tL}h\rangle + ct^3\|Ce^{-tL}h\|^2.$$

Then, we can perform computations similar to the ones in Subsection 4.2, except that now there are extra terms coming from the time-dependence of the coefficients a, b, c. Writing h for $e^{-tL}h$, we have, if $a, b/a, c/b, c^2/b, b^2/ac$ are small enough:

(A.21.19) $$\frac{d\mathcal{F}(t,h)}{dt} \leq -\kappa\Big(\|Ah\|^2 + at\|A^2h\|^2 + bt^2\|Ch\|^2 + ct^3\|CAh\|^2\Big)$$
$$+ a\|Ah\|^2 + 4bt\langle Ah, Ch\rangle + 3ct^2\|Ch\|^2,$$

were κ is a positive number. When $0 \leq t \leq 1$, the positive terms in the right-hand side of (A.21.19) can all be controlled by the negative terms if a, b and c/b are small enough. Then

$$\frac{d\mathcal{F}(t,h)}{dt} \leq -K(\|Ah\|^2 + \|A^2h\|^2 + \|Ch\|^2).$$

In particular, $\mathcal{F}$ is a nonincreasing function of t, and then the desired bounds $\|Ah\|^2 = O(t^{-1})$, $\|Ch\|^2 = O(t^{-3})$ follow (as well as the bound $\|A^2h\|^2 = O(t^{-2})$).

The very same scheme of proof allows to establish a regularization theorem similar to Theorem 24:

THEOREM A.12. *Let $\mathcal{H}$ be a Hilbert space, $A : \mathcal{H} \to \mathcal{H}^n$ and $B : \mathcal{H} \to \mathcal{H}$ be unbounded operators, $B^* = -B$, let $L := A^*A + B$. Assume the existence of $N_c \in \mathbb{N}$ and (possibly unbounded) operators $C_0, C_1, \ldots, C_{N_c+1}, R_1, \ldots, R_{N_c+1}$ and $Z_1, \ldots, Z_{N_c+1}$ such that*

$$C_0 = A, \qquad [C_j, B] = Z_{j+1}C_{j+1} + R_{j+1} \quad (0 \leq j \leq N_c), \qquad C_{N_c+1} = 0,$$

and, for all $k \in \{0, \ldots, N_c\}$,

(i) $[A, C_k]$ *is bounded relatively to* $\{C_j\}_{0 \leq j \leq k}$ *and* $\{C_j A\}_{0 \leq j \leq k-1}$;
(ii) $[C_k, A^*]$ *is bounded relatively to I and* $\{C_j\}_{0 \leq j \leq k}$;

(iii) R_k *is bounded relatively to* $\{C_j\}_{0\leq j\leq k-1}$ *and* $\{C_j A\}_{0\leq j\leq k-1}$;
(iv) There are positive constants λ_j, Λ_j *such that* $\lambda_j I \leq Z_j \leq \Lambda_j I$.
Then the following bound holds true along the semigroup e^{-tL}:

$$\forall k \in \{0,\ldots,N_c\}, \quad \forall t \in (0,1], \quad \|e^{-tL}h\| \leq C_k \frac{\|h\|}{t^{k+\frac{1}{2}}},$$

where C_k *only depends on the constants appearing implicitly in Assumptions (i)–(iv).*

REMARK A.13. A reasoning similar to Remark A.10 shows that the exponents $1/(k+1/2)$ cannot be improved. Indeed, in Hörmander's theory, the weight attributed to the commutator C_k would be $2k+1$, and the regularity estimates established by Rothschild and Stein [49], in general optimal, provide regularization by an order $2/(2k+1) = 1/(k+1/2)$.

REMARK A.14. I shall show below how Hérau's method can be adapted to yield regularization from $L \log L$ initial datum. On the other hand, it is not clear that it can be used to establish regularization from measure initial data.

A.21.3. Higher regularity from measure initial data. Now I shall explain how to extend the previous results by (a) establishing Sobolev regularity of higher order, (b) removing the assumption of L^2 integrability for the initial datum.

I shall only consider the case when ∇V is Lipschitz, for two reasons: (ii) this ensures the uniqueness of the solution of the Fokker–Planck equation starting from a *measure* initial datum; (ii) the theorems of convergence to equilibrium studied in the present paper use the Lipschitz regularity of ∇V anyway. To simplify the presentation I shall also assume that all derivatives of ∇V are uniformly (in x) bounded, even if in fact one only needs a finite number of such bounds, depending on the degree of regularity that one is aiming at.

As before, the equation under study is

$$(A.21.20) \qquad \partial_t f + v \cdot \nabla_x f - \nabla V(x) \cdot \nabla_v f = \Delta_v f + v \cdot \nabla_v f + nf.$$

This equation admits a unique solution as soon as f_0 is a finite nonnegative measure (say a probability measure) with finite energy, and it is easy to prove the propagation of regularity and of moment bounds.

So to establish regularization in higher-order Sobolev space $H_x^k(H_v^\ell)$ it is enough to prove, for smooth and rapidly decaying solutions, an a priori estimate like

$$\|f(t,\cdot)\|_{H_x^k H_v^\ell(\mathbb{R}_x^n \times \mathbb{R}_v^n)} \leq \frac{C}{t^\kappa}$$

with constants C and κ that do not depend on the regularity of f_0. Here is a precise result:

THEOREM A.15 (regularization from measure initial data). *Let* $\mu(dx\,dv)$ *be a probability measure on* $\mathbb{R}^n \times \mathbb{R}^n$, *and let* $V \in C^\infty(\mathbb{R}^n)$ *such that* $\sup |\nabla^k V(x)| < +\infty$ *for all* $k \geq 2$. *Then there is a unique solution* $f(t,x,v)$ *of (A.21.20), such that the probability measure* $\mu_t(dx\,dv) = f(t,x,v)\,dx\,dv$ *belongs to* $C(\mathbb{R}_+; P(\mathbb{R}^n \times \mathbb{R}^n))$ *and* $\lim_{t\to 0} \mu_t = \mu$. *Moreover, for any* $m \in \mathbb{N}$ *there are* $j = j(m,n) \in \mathbb{N}$, *and positive constants* $\kappa = \kappa(m,n)$, $C = C(m,n, \|\nabla^2 V\|_{C^j})$ *such that*

$$(A.21.21) \qquad 0 < t < 1 \Longrightarrow \sum_{k+3\ell \leq m} \|f(t,\cdot)\|_{H_x^k H_v^\ell} \leq \frac{C}{t^\kappa}.$$

PROOF OF THEOREM A.15. The solution of (A.21.20) can be seen as the law at time t of a stochastic process solving the stochastic differential equation $dx_t = v_t\, dt$, $dv_t = \sqrt{2}\, dW_t - v_t\, dt - \nabla V(x_t)\, dt$, which by assumption has uniformly Lipschitz coefficients. It is easy to deduce the existence and uniqueness of the solution. Then it is sufficient to establish (A.21.21) as an a priori estimate on C^∞ solutions.

As usual, C and K will stand for various constants depending only on n and V. As in Subsection A.21.1 the a priori estimate is divided into four steps. The conservation of mass (that is, the preservation of $\int f\, dx\, dv$) along equation (A.21.20) will be used several times.

Step 1: *"Energy" estimate in higher order Sobolev spaces.*

Let k and ℓ be given integers (k will be the regularity in x and ℓ the regularity in v). Computations similar to those in Subsection A.21.1 (differentiating the equation and integrating) yield

$$\frac{d}{dt}\int |\nabla_x^k \nabla_v^\ell f|^2\, dx\, dv \leq -K \int |\nabla_x^k \nabla_v^{\ell+1} f|^2\, dx\, dv$$
$$+ C \int |\nabla_x^k \nabla_v^\ell f|^2\, dx\, dv + C \int |\nabla_x^{k+1} \nabla_v^{\ell-2} f|^2\, dx\, dv$$
$$+ C \sum_{1 \leq i \leq k} \int |\nabla_x^{k-i} \nabla_v^\ell f|^2 |\nabla_x^{i+1} V|^2\, dx\, dv.$$

By assumption $|\nabla_x^i V|$ is bounded for any i, so the above equation reduces to

$$\frac{d}{dt}\int |\nabla_x^k \nabla_v^\ell f|^2 \leq -K \int |\nabla_x^k \nabla_v^{\ell+1} f|^2 + C \sum_{j \leq k} \int |\nabla_x^j \nabla_v^\ell f|^2$$
$$+ C \int |\nabla_x^{k+1} \nabla_v^{\ell-2} f|^2.$$

Then one can repeat the same computation with (k, ℓ) replaced by $(k+1, \ell-2)$ and then $(k+2, \ell-4)$, etc. By an easy induction, for any integer m, we can find positive constants $K, C, a_0 = 1, a_1, \ldots, a_m$ such that

$$\frac{d}{dt} \sum_{k=0}^m a_k \int |\nabla_x^k \nabla_v^{3(m-k)} f|^2\, dx\, dv \leq -K \sum_{k=0}^m \int |\nabla_x^k \nabla_v^{3(m-k)+1} f|^2\, dx\, dv$$
$$+ C \sum_{k=0}^m \sum_{j \leq k} \int |\nabla_x^j \nabla_v^{3(m-k)} f|^2\, dx\, dv.$$

Repeating the same operation for lower order terms (that is, decreasing m), for each couple of nonnegative integers (k, ℓ) with $3k + \ell \leq 3m$ we can find a positive constant $a_{k,\ell}$ such that

$$\frac{d}{dt} \sum_{3k+\ell \leq 3m} a_{k,\ell} \int |\nabla_x^k \nabla_v^\ell f|^2\, dx\, dv \leq -K \int |\nabla_v^{3m+1} f|^2\, dx\, dv$$
$$+ C \sum_{3k+\ell \leq 3m} \int |\nabla_x^k \nabla_v^\ell f|^2\, dx\, dv.$$

Then we can define an "energy functional" of order m, which controls the L^2-regularity of f up to order m in x and $3m$ in v:

$$\mathcal{E}_m(f) = \sum_{3k+\ell \leq 3m} a_{k,\ell} \int |\nabla_x^k \nabla_v^\ell f|^2 \, dx \, dv. \tag{A.21.22}$$

(Recall, to avoid any confusion, that $\int |\nabla_x^k \nabla_v^\ell f|^2$ is the sum of all terms $\int (\partial_{x_1}^{k_1} \ldots \partial_{x_n}^{k_n} \partial_{v_1}^{\ell_1} \ldots \partial_{v_n}^{\ell_n} f)^2$ with $k_1 + \ldots + k_n = k$, $\ell_1 + \ldots + \ell_n = \ell$.)

Then the a priori estimate on the Fokker–Planck equation (A.21.20) can be recast as

$$\frac{d}{dt} \mathcal{E}_m(f) \leq -K \int |\nabla_v^{3m+1} f|^2 \, dx \, dv + C \mathcal{E}_m(f). \tag{A.21.23}$$

The important terms in $\mathcal{E}_m$ are the extremal ones, that is for $(k, \ell) = (m, 0)$, $(0, 3m)$ or $(0, 0)$. All the other ones can be controlled by these three extremal terms; to see this, it suffices to apply Hölder's inequality in Fourier space: Denoting by ξ the conjugate variable to x and by η the conjugate variable to v, one has

$$\int |\nabla_x^k \nabla_v^\ell f|^2 \, dx \, dv = C \int |\xi|^{2k} |\eta|^{2\ell} |\widehat{f}| \, d\xi \, d\eta$$

$$\leq C \left(\int |\xi|^{2m} |\widehat{f}|^2 \, d\xi \, d\eta \right)^{\frac{k}{m}} \left(\int |\eta|^{6m} |\widehat{f}|^2 \, d\xi \, d\eta \right)^{\frac{\ell}{3m}}$$

$$\left(\int |\widehat{f}|^2 \, d\xi \, d\eta \right)^{1 - \left(\frac{k}{m} + \frac{\ell}{3m}\right)}$$

$$= C \left(\int |\nabla_x^m f|^2 \, dx \, dv \right)^{\frac{k}{m}} \left(\int |\nabla_v^{3m} f|^2 \, dx \, dv \right)^{\frac{\ell}{3m}}$$

$$\left(\int f^2 \, dx \, dv \right)^{1 - \left(\frac{k}{m} + \frac{\ell}{3m}\right)}.$$

It follows easily that there are positive constants K, C such that

$$K \left(\int |\nabla_x^m f|^2 + \int |\nabla_v^{3m} f|^2 + \int f^2 \right) \leq \mathcal{E}_m(f) \tag{A.21.24}$$

$$\leq C \left(\int |\nabla_x^m f|^2 + \int |\nabla_v^{3m} f|^2 + \int f^2 \right).$$

Step 2: *Mixed derivatives*

Define the higher order mixed derivative functional

$$\mathcal{M}_m(f) = \int \nabla_x^m f \cdot \nabla_x^{m-1} \nabla_v f \tag{A.21.25}$$

$$= \sum_{1 \leq i_1, \ldots, i_m \leq n} \int \frac{\partial^m f}{\partial x_{i_1} \ldots \partial x_{i_m}} \frac{\partial^m f}{\partial x_{i_1} \ldots \partial x_{i_{m-1}} \partial v_{i_m}}.$$

By computations in the same style as in Step 2 of Subsection A.21.1, one can establish

$$\frac{d}{dt} \mathcal{M}_m(f) \leq -K \int |\nabla_x^m f|^2 \, dx \, dv + C \sum_{k < m, \ 3k+\ell \leq 3m} \int |\nabla_x^k \nabla_v^\ell f|^2.$$

Each of the terms appearing in the latter sum can then be estimated by elementary interpolation inequalities as in Step 1: If $k < m$ then

$$\int |\nabla_x^k \nabla_v^\ell f|^2 \le \varepsilon \int |\nabla_x^m f|^2 + C \left(\int |\nabla_v^{3m} f|^2 + \int f^2 \right),$$

where ε is an arbitrarily small positive number. The conclusion is that
(A.21.26)
$$\frac{d}{dt} \mathcal{M}_m(f) \le -K \int |\nabla_x^m f|^2 \, dx \, dv + C \left(\int |\nabla_v^{3m} f|^2 \, dx \, dv + \int f^2 \, dx \, dv \right).$$

Step 3: *Interpolation inequalities.*

There are two things to check: (i) that $\mathcal{M}_m$ is "much smaller" than $\mathcal{E}_m$, and (ii) that $\int |\nabla_v^{3m} f|^2$ is "much smaller" than $\int |\nabla_v^{3m+1} f|^2$. The difficulty is that we cannot just use interpolation in L^2-type spaces. In replacement, we shall use the anisotropic *Nash*-type interpolation inequality exposed in Appendix A.23.

First, by Cauchy–Schwarz,

$$|\mathcal{M}_m(f)| \le \left(\int |\nabla_x^m f|^2 \right)^{\frac{1}{2}} \left(\int |\nabla_x^{m-1} \nabla_v f|^2 \right)^{\frac{1}{2}}.$$

Then the second term is estimated thanks to Lemma A.25 with $\lambda = m-1$, $\mu = 1$, $\lambda' = m$, $\mu' = 3m$:

$$\int |\nabla_x^{m-1} \nabla_v f|^2 \le \left(\int |\nabla_x^m f|^2 + \int |\nabla_v^{3m} f|^2 \right)^{1-\theta} \left(\int f \right)^{2\theta},$$

where $\theta = 2/(3m+6n)$ is a positive number. Since the mass $\int f$ is preserved under the time-evolution by the Fokker–Planck equation, we arrive at the estimate
(A.21.27)
$$|\mathcal{M}_m(f)| \le C \mathcal{E}_m(f)^{1-\delta},$$

where $\delta = \theta/2$ is a positive constant.

Next, apply Lemma A.25 again with $\lambda = 0$, $\lambda' = m$, $\mu = 3m$, $\mu' = 3m+1$. Noting that $(\lambda/\lambda') + (\mu/\mu') = 3m/(3m+1) < 1$, we see that there exists $\theta \in (0,1)$ such that

$$\int |\nabla_v^{3m} f|^2 \, dx \, dv$$
$$\le C \left(\int |\nabla_x^m f|^2 + \int |\nabla_v^{3m+1} f|^2 \, dx \, dv \right)^{1-\theta} \left(\int f \, dx \, dv \right)^{2\theta}.$$

The same estimate holds true for $\int f^2$ (this can be treated by the usual Nash inequality), and then one can use the fact that $\int f$ is preserved by the Fokker–Planck equation, to obtain the a priori estimate
(A.21.28)
$$\int |\nabla_v^{3m} f|^2 \, dx \, dv + \int f^2 \, dx \, dv \le C \left(\int |\nabla_x^m f|^2 + \int |\nabla_v^{3m+1} f|^2 \, dx \, dv \right)^{1-\theta}.$$

Step 4: *Conclusion*

Equations (A.21.24), (A.21.27), (A.21.23), (A.21.28) and (A.21.26) together show that we can apply Lemma A.26 with $\mathcal{E} = \mathcal{E}_m$, $\mathcal{M} = \mathcal{M}_m$, $X = \int |\nabla_x^m f|^2$, $Y = \int |\nabla_v^{3m} f|^2 + \int f^2$, $Z = \int |\nabla_v^{3m+1} f|^2$. Thus there are constants C and κ such that $\mathcal{E}_m(f_t) \le C/t^{1/\kappa}$. This concludes the proof of the a priori estimate. □

REMARK A.16. Keeping track of the constants in the above proof yields the bound $\|\nabla_x^m f(t)\|_{L^2}^2 + \|\nabla_v^{3m} f(t)\|^2 = O(t^{-(3m+6n)})$ for any $m \in \mathbb{N}$ ($m \geq 1$). This exponent is not optimal, and can be improved as follows: first apply the above bound with m replaced by $m' > m$, and then use inequality (A.25) to interpolate the norm of index m ($m \geq 0$, possibly noninteger) between the norm of index m' and the L^1 norm. The result is a bound like $O(t^{-\beta(n,m,m')})$ with $\beta(n,m,m') = (3m+2n)(1+(4n)/(3m'+2n))$, and the limit $m' \to \infty$ provides the following Sobolev estimate on the density at time t ($0 < t < 1$):

$$(A.21.29) \quad \forall m \geq 0 \qquad \|\nabla_x^m f(t)\|_{L^2}^2 + \|\nabla_v^{3m} f(t)\|^2 = O(t^{-\beta}), \qquad \forall \beta > 3m+2n.$$

The constant in the right-hand side may depend on β but not on the initial datum, which may be any probability measure. It is natural to conjecture that the optimal exponent in (A.21.29) is exactly $\beta_{m,n} = 3m + 2n$; in the simple case when the potential is quadratic this can be proven by a direct estimate on the fundamental solution. (Compare with the bound $\|\nabla_x^m f\|^2 + \|\nabla_v^m f\|^2 = O(t^{-(m+n)})$ for the usual heat equation in $\mathbb{R}^{2n}$).

REMARK A.17. The particular case $m = 0$ in (A.21.29) is of particular interest, since it corresponds to a uniform pointwise bound on the fundamental solution of the equation. In the context of symmetric operators, such pointwise bounds have been investigated with extreme care [**27, 51**], and are known to be very well encoded by Nash inequalities in the style of (A.23.4). For instance, for the heat equation in $\mathbb{R}^n$, the bound $\|f(t)\|_{L^2}^2 = O(t^{-\gamma})$ with $\gamma = n/2$ is obtained from $\theta = 1/(n/2+1)$ in (A.23.4) via $\gamma = (1-\theta)/\theta$. For the hypoelliptic diffusion which we are considering, one may ask whether the presumably optimal bound $\|f(t)\|_{L^2}^2 = O(t^{-2n})$ is captured by a functional inequality of Nash style. I don't know the answer to this question, but if it is affirmative then I claim that the functional inequality must be

$$(A.21.30) \quad \int f^2 \leq C \left(\int |\nabla_v f|^2 + \int |D_x^{1/2} D_v^{-1/2} f|^2 \right)^{1-\theta} \left(\int f \right)^{2\theta},$$

where all integrals are with respect to $dx\, dv$, and

$$\theta = \frac{1}{2n+1}.$$

The unorthodox Nash-type inequality (A.21.30) can be proven with the same technique which is used to prove Lemma A.25 later. But I have been unable to get the desired short-time estimate from (A.21.30)! A natural strategy was to look for a scalar product $\langle\,,\,\rangle$, defining a norm equivalent to the L^2 norm, such that $\langle Lf, f \rangle \geq K \left(\int |\nabla_v f|^2 + \int |D_x^{1/2} D_v^{-1/2} f|^2 \right)$, and then conclude by Gronwall's lemma; but it is not clear that such a scalar product exists (the natural candidate $\int f^2 + a \int (D_v^{-1} \nabla_v f) \cdot (D_x^{-1} \nabla_x f)$ does not seem to work). This is a borderline problem: if one replaces $D_v^{-1/2}$ by D_v^α with $\alpha > -1/2$, the program can be carried out, and yields a bound $O(t^{-\beta})$ for any $\beta > 2n$ (which was already obtained in Remark A.16 by a slightly different, although closely related, method).

A.21.4. Regularization in an $L \log L$ context. If the initial datum is assumed to have finite entropy, then Hérau's method can be adapted to yield the regularization in Fisher information sense, with exponents that are likely to be optimal. Here is a rather general result in this direction, under the same assumptions as Theorem 28:

THEOREM A.18. *Let $E \in C^2(\mathbb{R}^N)$, such that e^{-E} is rapidly decreasing, and $\mu(dX) = e^{-E(X)}\, dX$ is a probability measure on $\mathbb{R}^N$. Let $(A_j)_{1 \leq j \leq m}$ and B be first-order derivation operators with smooth coefficients. Denote by A_j^* and B^* their respective adjoints in $L^2(\mu)$, and assume that $B^* = -B$. Denote by A the collection $(A_1, \ldots, A_m)$, viewed as an unbounded operators whose range is made of functions valued in $\mathbb{R}^m$. Define*

$$L = A^*A + B = \sum_{j=1}^m A_j^* A_j + B,$$

and assume that e^{-tL} defines a well-behaved semigroup on a suitable space of positive functions (for instance, $e^{-tL}h$ and $\log(e^{-tL}h)$ are C^∞ and all their derivatives grow at most polynomially if h is itself C^∞ with all derivatives bounded, and h is bounded below by a positive constant).

Next assume the existence of an integer $N_c \geq 1$, derivation operators $C_0, \ldots, C_{N_c+1}$ and $R_1, \ldots, R_{N_c+1}$, and vector-valued functions $Z_1, \ldots, Z_{N_c+1}$ (all of them with C^∞ coefficients, growing at most polynomially, as their partial derivatives) such that

$$C_0 = A, \qquad [C_j, B] = Z_{j+1}\, C_{j+1} + R_{j+1} \quad (0 \leq j \leq N_c), \qquad C_{N_c+1} = 0,$$

and

(i) $[A, C_k]$ is pointwise bounded relatively to A;
(ii) $[C_k, A^*]$ is pointwise bounded relatively to $I, \{C_j\}_{0 \leq j \leq k}$;
(iii) R_k is pointwise bounded with respect to $\{C_j\}_{0 \leq j \leq k-1}$;
(iv) there are positive constants λ_j, Λ_j such that $\lambda_j \leq Z_j \leq \Lambda_j$;
(v) $[A, C_k]^*$ is pointwise bounded relatively to I, A.
Then the following bound holds true: With the notation $h(t) = e^{-tL}h_0$,

$$\forall k \in \{0, \ldots, N_c\}, \quad \forall t \in (0, 1],$$

$$\int h(t) |C_k \log h(t)|^2\, d\mu \leq C_k \frac{\int h_0 \log h_0\, d\mu}{t^{2k+1}},$$

where C_k is a constant only depending on the constants appearing implicitly in Assumptions (i)–(v).

PROOF. The proof is patterned after the proofs of Theorems 28 and A.12: Write $u = \log h$, $f = e^{-E}h$, and introduce the Lyapunov functional

$$\mathcal{F}(t, h) = \int fu + \sum_{k=0}^{N_c} \left(a_k t^{2k+1} \int f |C_k u|_m^2 + 2 b_k t^{2k+2} \int f \langle C_k u, C_{k+1} u \rangle_m \right).$$

The computations for $d\mathcal{F}/dt$ are the same as in the proof of Theorem 28, except that now there are additional terms caused by the explicit dependence on t. So

$$(A.21.31) \quad \frac{d\mathcal{F}(t,h(t))}{dt} \leq$$

$$-K\left(\int f|Au|^2 + \sum_k a_k t^{2k+1}\int f|C_k Au|^2 + \sum_k b_k t^{2k+2}\int f|C_{k+1}u|^2\right)$$

$$+\sum_k (2k+1)a_k t^{2k}\int f|C_k u|^2 + \sum_k (2k+2)b_k t^{2k+1}\int f\langle C_k u, C_{k+1}u\rangle.$$

Obviously, the additional terms can be controlled by the ones in the first line of the right-hand side, provided that the ratios a_k/b_{k-1} and $b_k t^{2k+1}/\sqrt{(b_{k-1}t^{2k})(b_k t^{2k+1})}$ are small enough; the second condition reduces to b_k/b_{k-1} small enough. These conditions have been enforced in the proof of Theorem 28. So all in all, $\mathcal{F}(t,h(t))$ is a nonincreasing function of t, and the conclusion follows immediately. □

A.22. Local positivity estimates

In this Appendix I shall derive local (strict) positivity estimates for smooth solutions of linear equations of Fokker–Planck type. Such estimates are a classical topic for hypoelliptic equations in general, and are often used in probability theory; in this memoir they were useful for the proof of Theorem 56. The results established below are probably not new, but the method is quite flexible and elementary, relying mainly on the maximum principle. As a main shortcoming, the positivity estimate will use smoothness, so it does not behave well in very short time; still it is quite sufficient for our purposes.

In the sequel, I shall use the notation

$$B_r(x_0, v_0) = \left\{(x,v) \in \mathbb{R}^n \times \mathbb{R}^n;\ |v - v_0| \leq r,\ |x - x_0| \leq r^3\right\}.$$

THEOREM A.19 (spreading of positivity). *Let $f = f(t,x,v)$ be a classical nonnegative solution of*

$$(A.22.1) \quad \frac{\partial f}{\partial t} + v \cdot \nabla_x f - \Delta_v f = A(t,x,v) \cdot \nabla_v f + B(t,x,v)\,f$$

in $[0,T] \times \Omega$, where Ω is an open subset of $\mathbb{R}^n_x \times \mathbb{R}^n_v$, and $A: [0,T] \times \Omega \to \mathbb{R}^n$, $B: [0,T] \times \Omega \to \mathbb{R}$ are bounded continuous functions. Let $(x_0, v_0) \in \Omega$, $V \geq |v_0|$, $\overline{A} \geq \|A\|_{L^\infty}$, $\overline{B} \geq \|B\|_{L^\infty}$. Then for any $r, \tau > 0$ there are constants $\lambda, K > 0$, only depending on n, $\overline{A}$, $\overline{B}$ and r^2/τ, with the following property: If $B_{\lambda r}(x_0, v_0) \subset \Omega$, $\tau \leq \min(1, T, r^3/(4V))$ and $f \geq \delta > 0$ in $[0,\tau) \times B_r(x_0, v_0)$, then $f \geq K\delta$ in $[\tau/2, \tau) \times B_{2r}(x_0, v_0)$.

Theorem A.19 implies, via covering arguments in all variables t, x, v:

COROLLARY A.20. *If $f \geq 0$ solves (A.22.1) in $[0,T) \times \Omega$ and $f \geq \delta > 0$ in $[0,T) \times B_r(x_0, v_0)$, then for any compact set $K \subset \Omega$ containing (x_0, v_0) and for any $t_0 \in (0,T)$, we have $f \geq \delta' > 0$ in $[t_0, T) \times K$, where δ' only depends on $n, \overline{A}, \overline{B}, K, \Omega, x_0, v_0, r, t_0, \delta$.*

Now assume that f has mass at least $m > 0$ in some compact subset K of Ω. Then for each $t_0 > 0$ there is some $(x_0, v_0) \in K$ such that $f(t_0, x_0, v_0) \geq m/|K|$, where $|K|$ stands for the Lebesgue measure of K. If moreover f is L-Lipschitz in (t,x,v) then there are $r > 0$ and $\tau > 0$, controlled from below, such that $f \geq m/(2|K|)$ on $[t_0, t_0 + \tau] \times B_r(x_0, v_0)$. Combining this with Corollary A.20 and a covering argument again, we arrive at the following useful result ($\|f\|_{\text{Lip}}$ stands for the Lipschitz norm of f):

COROLLARY A.21 (uniform local lower bounds in positive time). *If $f \geq 0$ solves (A.22.1) in $[0,T) \times \Omega$ and K is a compact subset of Ω such that (i) $\int_K f\, dx\, dv \geq m > 0$, (ii) $\|f\|_{\text{Lip}([0,T] \times O)} \leq L$ for some neighborhood O of K in Ω, then for any $t_0 > 0$ there is $\delta = \delta(n, \overline{A}, \overline{B}, K, O, \Omega, m, L, t_0) > 0$ such that $f \geq \delta$ on $[t_0, T) \times K$.*

In the end the Lipschitz assumption in the above corollary can of course be removed by a hypoellipticity argument. Now let us prove Theorem A.19. The argument is based on the maximum principle in the style of [**13**, Section 10], together with an "algebraic trick" of the kind which I have used at various places in this memoir (and a bit of juggling with parameters).

PROOF OF THEOREM A.19. Let $g(t,x,v) = e^{\overline{B}t} f(t,x,v)$; then $g \geq f$ and $\mathcal{L}g \geq 0$ in $(0,T) \times \Omega$, where

$$\mathcal{L} = \frac{\partial}{\partial t} + v \cdot \nabla_x - \Delta_v - A(t,x,v) \cdot \nabla_v.$$

Let us construct a particular subsolution for $\mathcal{L}$. In the sequel, B_r will stand for $B_r(x_0, v_0)$. For $t \in (0, \tau]$ and $(x, v) \in \Omega \setminus B_r$ let

$$Q(t,x,v) = a\frac{|v - v_0|^2}{2t} - b\frac{\langle v - v_0, x - X_t(x_0, v_0)\rangle}{t^2} + c\frac{|x - X_t(x_0, v_0)|^2}{2t^3},$$

where $X_t(x_0, v_0) = x_0 + tv_0$ (abbreviated X_t in the sequel) is the position at time t of the geodesic flow starting from (x_0, v_0), and $a, b, c > 0$ will be chosen later on. Let further

$$\varphi(t,x,v) = \delta\, e^{-\mu Q(t,x,v)} - \varepsilon,$$

where $\mu, \varepsilon > 0$ will be chosen later on. Let us assume $b^2 < ac$, so that Q is a positive definite quadratic form in the two variables $v - v_0$ and $x - X_t$. Then

$$\mathcal{L}\varphi = \mu\, \delta\, e^{-\mu Q}\, \mathcal{A}(Q),$$

where

$$\mathcal{A}(Q) = \partial_t Q + v \cdot \nabla_x Q - \Delta_v Q + \mu |\nabla_v Q|^2 - A(t,x,v) \cdot \nabla_v Q.$$

By computation,

$$\begin{aligned}\mathcal{A}(Q) = &-a\frac{|v-v_0|^2}{2t^3} + 2b\frac{\langle v-v_0, x-X_t\rangle}{t^3} - b\frac{|v-v_0|^2}{t^2} \\ &- 3c\frac{|x-X_t|^2}{2t^4} + c\frac{\langle x-X_t, v-v_0\rangle}{t^3} - a\frac{n}{t} \\ &+ \mu\left|a\frac{(v-v_0)}{t} - b\frac{(x-X_t)}{t^2}\right|^2 - a\frac{\langle A, v-v_0\rangle}{t} + b\frac{\langle A, x-X_t\rangle}{t^2} \\ =\; &\mathcal{B}\left(\frac{v-v_0}{t}, \frac{x-X_t}{t^2}\right) - a\frac{\langle A, v-v_0\rangle}{t} + b\frac{\langle A, x-X_t\rangle}{t^2} - a\frac{n}{t},\end{aligned}$$

where $\mathcal{B}$ is a quadratic form on $\mathbb{R}^n \times \mathbb{R}^n$ with matrix $M \otimes I_n$,

$$M = \begin{bmatrix} \mu a^2 - \dfrac{a}{2} - b & b + \dfrac{c}{2} - \mu ab \\ b + \dfrac{c}{2} - \mu ab & \mu b^2 - 3c \end{bmatrix}.$$

If a, b, c are given, then as $\mu \to \infty$,

$$\begin{cases} \operatorname{tr} M = \mu(a^2 + b^2) + O(1), \\ \det M = \mu \left[\dfrac{5}{2} ab^2 + abc - b^3 - 3a^2 c \right] + O(1). \end{cases}$$

Both quantities are positive if $b \gg a$ and $ac \gg b^2$; then as $\mu \to \infty$ the eigenvalues of M are of order μb^2 and $ac/b \gg b$. So for any fixed C we may choose a, b, c and μ so that

$$\mathcal{B}\left(\frac{v - v_0}{t}, \frac{x - X_t}{t^2} \right) \geq C b \left(\frac{|v - v_0|^2}{t^2} + \frac{|x - X_t|^2}{t^4} \right).$$

Combining this with the obvious bounds $|\langle A, (v - v_0)/t \rangle| \leq (\overline{A})^2/2 + |v - v_0|^2/(2t^2)$ and $|\langle A, (x - X_t)/t^2 \rangle| \leq (\overline{A})^2 + |x - X_t|^2/(2t^4)$, assuming $\tau \leq 1$, we get in the end

$$\mathcal{A}(Q) \geq \text{const.} \frac{b}{t} \left[C \left(\frac{|v - v_0|^2}{t} + \frac{|x - X_t|^2}{t^3} \right) - 1 \right],$$

where C is arbitrarily large.

Recall that $(x, v) \notin B_r$; so
- either $|v - v_0| \geq r$, and then $\mathcal{A}(Q) \geq \text{const.}\,(b/t)[Cr^2/\tau - 1]$, which is positive if $C > \tau/r^2$;
- or $|x - x_0| \geq r^3$, and then, for any $\tau \leq r^3/(4V)$,

$$\frac{|x - X_t|^2}{t^2} \geq \frac{|x - x_0|^2}{2t^2} - 2|v_0|^2 \geq \frac{r^6}{2\tau^2} - 2V^2 \geq \frac{r^6}{4\tau^2};$$

so $\mathcal{A}(Q) \geq \text{const.}\,(b/t)[Cr^6/4\tau^3 - 1]$, which is positive as soon as $C > 4(\tau/r^2)^3$.

To summarize: under our assumptions there is a way to choose the constants a, b, c, μ, depending only on $n, \overline{A}, r^2/\tau$, satisfying $c \gg b \gg a \gg 1$ and $bc \gg a^2$, so that $\mathcal{L}\varphi \leq 0$ in $[0, \tau) \times (B_{\lambda r} \setminus B_r)$, as soon as $\tau \leq \min(1, T, r^3/(4V))$. We now wish to enforce $\varphi \leq g$ for $t = 0$ and for $(x, v) \in \partial(B_{\lambda r} \setminus B_r)$; then the classical maximum principle will imply $g \geq \varphi$ in $[0, \tau) \times (B_{\lambda r} \setminus B_r)$.

The boundary condition at $t = 0$ is obvious since φ vanishes identically there (more rigorously, φ can be extended by continuity by 0 at $t = 0$). The condition is also true on ∂B_r since $\varphi \leq \delta$ and $g \geq \delta$. It remains to impose it on $\partial B_{\lambda r}$. For that we estimate Q from below: as soon as ac/b^2 is large enough,

$$Q(t, x, v) \geq \frac{a}{4} \left(\frac{|v - v_0|^2}{t} + \frac{|x - X_t|^2}{t^3} \right),$$

so for $(x, v) \in \partial B_{\lambda r}$ a computation similar to the one above yields

$$Q(t, x, v) \geq \frac{a}{4} \min\left(\frac{\lambda^2 r^2}{\tau}, \frac{\lambda^6 r^6}{4\tau^3} \right) \geq \frac{a \lambda^2}{16} \min\left(\frac{r^2}{\tau}, \left(\frac{r^2}{\tau} \right)^3 \right).$$

Thus if we choose

$$\varepsilon = \delta \exp\left(-\frac{\mu a \lambda^2}{16} \min\left(\frac{r^2}{\tau}, \left(\frac{r^2}{\tau}\right)^3\right)\right),$$

we make sure that $\varphi = \delta e^{-\mu Q} - \varepsilon \leq 0$ on $\partial B_{\lambda r}$, a fortiori $\varphi \leq g$ on this set, and then we can apply the maximum principle.

So now we have $\varphi \leq g$, and this will yield a lower bound for g in $[\tau/2, \tau) \times (B_{2r} \setminus B_r)$: indeed, if $t \geq \tau/2$ and $(x, v) \in B_{2r} \setminus B_r$ then

$$Q(t, x, v) \leq 2c \left(\frac{|v - v_0|^2}{t} + \frac{|x - X_t|^2}{t^3}\right) \leq 64\, c \left(\frac{r^2}{\tau} + \frac{r^6}{\tau^3} + \frac{V^2}{\tau}\right)$$

$$\leq 160\, c \max\left(\frac{r^2}{\tau}, \left(\frac{r^2}{\tau}\right)^3\right);$$

so

$$\varphi(t, x, v) \geq \delta \left[\exp\left(-160\, \mu\, c \max\left(\frac{r^2}{\tau}, \left(\frac{r^2}{\tau}\right)^3\right)\right)\right.$$

$$\left. - \exp\left(-\frac{\mu a \lambda^2}{16} \min\left(\frac{r^2}{\tau}, \left(\frac{r^2}{\tau}\right)^3\right)\right)\right].$$

For λ large enough the right-hand side is bounded below by $K_0 \delta$, where $K_0 = \exp\left(-160\, \mu\, c \max\left(\frac{r^2}{\tau}, \left(\frac{r^2}{\tau}\right)^3\right)\right)/2$; so g is bounded below by $K_0 \delta$, and $f \geq (K_0 \exp(-\tau \overline{B}))\delta$ on $[\tau/2, \tau) \times (B_{2r} \setminus B_r)$. This completes the proof. $\square$

A.23. Toolbox

The following elementary lemma is used in the proofs of Theorems 18, 42 and 27.

LEMMA A.22. *Let $\delta > 0$ and $u_0 > 0$ be given. Then it is always possible to choose positive numbers $u_1, u_2, \ldots, u_N$ in such a way that*

$$\begin{cases} \forall k \in \{0, \ldots, N-1\}, & u_{k+1} \leq \delta\, u_k; \\ \forall k \in \{1, \ldots, N-1\}, & u_k^2 \leq \delta\, u_{k-1}\, u_{k+1}. \end{cases}$$

PROOF. Without loss of generality, assume $u_0 = 1$. Set $m_0 = 0, m_1 = 1$; by induction, it is possible to pick up positive numbers m_k such that

$$m_{k+1} \in (m_k, 2m_k - m_{k-1}).$$

The resulting sequence will be increasing and satisfy $m_k > (m_{k-1} + m_{k+1})/2$. Next set $u_k = \varepsilon^{m_k}$; for ε small enough, the desired inequalities are satisfied. $\square$

The next lemma is used in the proof of Theorem 52; it is a kind of nonlinear analogue of Lemma A.22.

LEMMA A.23. *Let $K, \overline{E}, k > 0, J$ be given. Then there exist constants $\varepsilon_1 = \varepsilon_1(J) > 0$, $\ell = \ell(J, k) > 0$ and $K_1 = K_1(K, \overline{E}, k, J) > 0$ with the following*

A.23. TOOLBOX

property: For any $\varepsilon \in (0, \varepsilon_1)$ and $E \in (0, \overline{E})$, there exist coefficients $a_1, \ldots, a_{J-1} > 0$ satisfying

(A.23.1)
$$\begin{cases} 1 = a_0 \geq a_1 \geq a_2 \geq \ldots \geq a_{J-1}; \\ a_1 \leq K E^\varepsilon; \\ \forall j \in \{1, J-1\}, \quad \dfrac{a_j^2}{a_{j-1}} \leq K a_{J-1}^{1+\varepsilon} E^{k\varepsilon}; \\ a_{J-1} \geq K_1 E^{\ell \varepsilon}. \end{cases}$$

PROOF OF LEMMA A.23. Without loss of generality we may assume that K is bounded above by $m := \min(1, (\overline{E})^{-k})$; otherwise, just replace K by m.

We shall choose the coefficients a_j in such a way that the inequality in the third line of (A.23.1) holds as an equality. For $j = J-1$ this gives
$$a_{J-1}^2 = K a_{J-1}^{1+\varepsilon} a_{J-2} E^{k\varepsilon},$$
hence
$$a_{J-2} = a_{J-1} (K E^{k\varepsilon} a_{J-1}^\varepsilon)^{-1}.$$
Then the equality
$$\frac{a_j}{a_{J-1}} = \left(\frac{a_{j+1}}{a_{J-1}}\right)^2 (K E^{k\varepsilon} a_{J-1}^\varepsilon)^{-1}$$
yields, by decreasing induction,
$$a_j = a_{J-1} (K E^{k\varepsilon} a_{J-1}^\varepsilon)^{-\alpha_j},$$
where α_j is defined by the (decreasing) recursion relation
$$\alpha_{J-2} = 1, \qquad \alpha_{j-1} = 2\alpha_j + 1.$$

The sequence $(a_j)_{1 \leq j \leq J-1}$ is nonincreasing if $K E^{k\varepsilon} a_{J-1} \leq 1$. From the bound $K \leq \min(1, (\overline{E})^{-k})$, we know that $K E^{k\varepsilon} \leq 1$ as soon as $\varepsilon \leq \varepsilon_1 \leq 1$ (ε_1 to be chosen later), and $a_{J-1} \leq 1$.

Then $\alpha_1 = 2^{J-2} - 1$ is a positive integer depending only on J, and
$$a_1 = a_{J-1} (K E^{k\varepsilon} a_{J-1}^\varepsilon)^{-\alpha_1} = a_{J-1}^{1-\alpha_1 \varepsilon} (K E^{k\varepsilon})^{-\alpha_1}.$$
If $\varepsilon \leq \varepsilon_1 := 1/(2\alpha_1)$, then a_{J-1} appears in the right-hand side with a positive exponent $1 - \alpha_1 \varepsilon \in (1/2, 1)$. Also $\varepsilon_1 \leq 1$, as assumed before.

To make sure that the first condition in (A.23.1) is fulfilled, we impose
$$a_{J-1}^{1-\alpha_1 \varepsilon} (K E^{k\varepsilon})^{-\alpha_1} = K E^\varepsilon,$$
that is
$$a_{J-1} = \left[K^{1+\alpha_1} E^{(1+k\alpha_1)\varepsilon}\right]^{\frac{1}{1-\alpha_1 \varepsilon}}.$$
Up to decreasing K again, we may assume that the quantity inside square brackets is bounded by 1; this also implies that $a_{J-1} \leq 1$, as assumed before. Then, since $1/(1-\alpha_1 \varepsilon) \leq 1/(1-\alpha_1 \varepsilon_1) = 2$, one has
$$a_{J-1} \geq \left[K^{1+\alpha_1} E^{(1+k\alpha_1)\varepsilon}\right]^2,$$
and the lemma follows upon choosing $K_1 = K^{2(1+\alpha_1)}$, $\ell = 2(1+k\alpha_1)$. □

The next lemma, used to check (7.2) in Section 7, states that $|\nabla^2 V|$ defines a bounded operator $H^1(e^{-V}) \to L^2(e^{-V})$ as soon as $|\nabla^2 V|$ is dominated by $|\nabla V|$.

LEMMA A.24. *Let V be a C^2 function on $\mathbb{R}^n$, satisfying (7.3). Then, for all $g \in H^1(e^{-V})$,*

(i) $\displaystyle\int_{\mathbb{R}^n} |\nabla V|^2 g^2 e^{-V} \leq 8(1+\sqrt{n}C)^2 \left(\int_{\mathbb{R}^n} g^2 e^{-V} + \int_{\mathbb{R}^n} |\nabla g|^2 e^{-V} \right);$

(ii) $\displaystyle\int_{\mathbb{R}^n} |\nabla^2 V|^2 g^2 e^{-V} \leq 16\, C^2(1+\sqrt{2n}C)^2 \left(\int_{\mathbb{R}^n} g^2 e^{-V} + \int_{\mathbb{R}^n} |\nabla g|^2 e^{-V} \right).$

PROOF OF LEMMA A.24. By a density argument, we may assume that g is smooth and decays fast enough at infinity. Then, by the identity $\nabla(e^{-V}) = -(\nabla V)e^{-V}$ and an integration by parts,

$$\int |\nabla V|^2 g^2 e^{-V} = -\int g^2 \nabla V \cdot \nabla(e^{-V}) = \int \nabla \cdot (g^2 \nabla V)\, e^{-V}$$
$$= \int g^2 (\Delta V) e^{-V} + 2\int g(\nabla g \cdot \nabla V)\, e^{-V}.$$

By Cauchy–Schwarz inequality,

$$(A.23.2) \quad \int |\nabla V|^2 g^2 e^{-V} \leq \sqrt{\int g^2 (\Delta V)^2 e^{-V}} \sqrt{\int g^2 e^{-V}}$$
$$+ 2\sqrt{\int |\nabla V|^2 g^2 e^{-V}} \sqrt{\int |\nabla g|^2 e^{-V}}.$$

Since, by (7.5),

$$(\Delta V)^2 \leq n|\nabla^2 V|^2 \leq nC^2(1+|\nabla V|)^2 \leq 2nC^2(1+|\nabla V|^2),$$

it follows from (A.23.2) that

$$\int |\nabla V|^2 g^2 e^{-V} \leq \sqrt{2n}\,C\, \sqrt{\int g^2 e^{-V} + \int |\nabla V|^2 g^2 e^{-V}} \sqrt{\int g^2 e^{-V}}$$
$$+ 2\sqrt{\int |\nabla V|^2 g^2 e^{-V}} \sqrt{\int |\nabla g|^2 e^{-V}}$$
$$\leq \sqrt{2n}\,C \int g^2 e^{-V} + \sqrt{2n}\,C \sqrt{\int |\nabla V|^2 g^2 e^{-V}} \sqrt{\int g^2 e^{-V}}$$
$$+ 2\sqrt{\int |\nabla V|^2 g^2 e^{-V}} \sqrt{\int |\nabla g|^2 e^{-V}}.$$

Thanks to Young's inequality, this can be bounded by

$$\sqrt{2n}\,C \int g^2 e^{-V} + \left(\frac{1}{4}\int |\nabla V|^2 g^2 e^{-V} + 2nC^2 \int g^2 e^{-V}\right)$$
$$+ \left(\frac{1}{4}\int |\nabla V|^2 g^2 e^{-V} + 4\int |\nabla g|^2 e^{-V}\right).$$

All in all,

$$\int |\nabla V|^2 g^2\, e^{-V} \le \frac{1}{2} \int |\nabla V|^2 g^2\, e^{-V} + (\sqrt{2n}\,C + 2nC^2) \int g^2\, e^{-V}$$
$$+ 4 \int |\nabla g|^2\, e^{-V},$$

so

(A.23.3) $$\int |\nabla V|^2 g^2\, e^{-V} \le 2(\sqrt{2n}\,C + 2nC^2) \int g^2\, e^{-V} + 8 \int |\nabla g|^2\, e^{-V},$$

This easily leads to statement (i) after crude upper bounds.

To prove statement (ii), start again from (A.23.3) and apply (7.3) again, in the form $|\nabla^2 V|^2 \le 2C(1 + |\nabla V|^2)$: the desired conclusion follows at once. □

Next, we shall study an interpolation inequality "in Nash style". First recall the classical Nash inequality [44] in $\mathbb{R}^n_x$: If f is a nonnegative function of $x \in \mathbb{R}^n$, then

(A.23.4) $$\int_{\mathbb{R}^n} f^2\, dx \le C(n) \left(\int_{\mathbb{R}^n} |\nabla_x f|^2\, dx \right)^{1-\theta} \left(\int_{\mathbb{R}^n} f\, dx \right)^{2\theta},$$

where

$$\theta = \frac{2}{n+2}.$$

It is easy to generalize this inequality to higher orders, or fractional derivatives: If $D = (-\Delta)^{1/2}$, and $0 \le \lambda < \lambda'$, then

$$\int_{\mathbb{R}^n} |D_x^\lambda f|^2\, dx \le C(n, \lambda, \lambda') \left(\int_{\mathbb{R}^n} |D_x^{\lambda'} f|^2\, dx \right)^{1-\theta} \left(\int_{\mathbb{R}^n} f\, dx \right)^{2\theta},$$

where now

$$\theta = \frac{2(\lambda' - \lambda)}{n + 2\lambda'}.$$

The next lemma generalizes this to functions which depend on two variables, x and v, and allows different orders of derivations in these variables. The symbol D will again stand for $(-\Delta)^{1/2}$.

LEMMA A.25. *Let $f = f(x,v)$ be a nonnegative (smooth, rapidly decaying) function on $\mathbb{R}^n_x \times \mathbb{R}^n_v$. Let $\lambda, \lambda', \mu, \mu'$ be four nonnegative numbers with $\lambda', \mu' > 0$. If*

$$\frac{\lambda}{\lambda'} + \frac{\mu}{\mu'} < 1,$$

then there is a constant $C = C(n, \lambda, \mu, \lambda', \mu')$ such that

(A.23.5) $$\int |D_x^\lambda D_v^\mu f|^2\, dx\, dv$$
$$\le C \left(\int |D_x^{\lambda'} f|^2\, dx\, dv + \int |D_v^{\mu'} f|^2\, dx\, dv \right)^{1-\theta} \left(\int f \right)^{2\theta},$$

where

$$\theta = \frac{1 - \left(\frac{\lambda}{\lambda'} + \frac{\mu}{\mu'} \right)}{1 + \frac{n}{2}\left(\frac{1}{\lambda'} + \frac{1}{\mu'} \right)}.$$

PROOF OF LEMMA A.25. The strategy here will be the same as in the classical proof (actually due to Stein) of Nash's inequality: Go to Fourier space and separate according to high and low frequencies, then optimize. I shall denote by $\widehat{f}$ the Fourier transform of f, by ξ the Fourier variable that is dual to x, and by η the variable that is dual to v. So the inequality to prove is

$$(A.23.6) \quad \int |\xi|^{2\lambda} |\eta|^{2\mu} |\widehat{f}|^2 \, d\xi \, d\eta$$

$$\leq C \left(\int |\xi|^{2\lambda'} |\widehat{f}|^2 \, d\xi \, d\eta + \int |\eta|^{2\mu'} |\widehat{f}|^2 \, d\xi \, d\eta \right)^{1-\theta} \|\widehat{f}\|_{L^\infty}^{2\theta}.$$

First start with the case $\lambda = 0$, and separate the integral in the left-hand side of (A.23.6) into three parts:

$$\int (\ldots) \, d\xi \, d\eta = \int_{|\xi| \leq R, \, |\eta| \leq S} (\ldots) \, d\xi \, d\eta + \int_{|\xi| > R, \, |\eta| \leq S} (\ldots) \, d\xi \, d\eta$$

$$+ \int_{|\xi| > R, \, |\eta| > S} (\ldots) \, d\xi \, d\eta,$$

where R and S are positive numbers that will be chosen later on.

Then,

$$\int_{|\xi| \leq R, \, |\eta| \leq S} |\eta|^{2\mu} |\widehat{f}(\xi, \eta)|^2 \, d\eta \, d\xi \leq S^{2\mu} \operatorname{vol}(|\xi| \leq R) \operatorname{vol}(|\eta| \leq S) \|\widehat{f}\|_{L^\infty}^2$$

$$(A.23.7) \qquad\qquad\qquad \leq C_n R^n S^{2\mu + n} \|\widehat{f}\|_{L^\infty}^2,$$

where C_n only depends on n, and vol is a notation for the Lebesgue volume in $\mathbb{R}^n$.

Next

$$(A.23.8) \quad \int_{|\xi| > R, \, |\eta| \leq S} |\eta|^{2\mu} |\widehat{f}(\xi, \eta)|^2 \, d\xi \, d\eta \leq \frac{S^{2\mu}}{R^{2\lambda'}} \int |\xi|^{2\lambda'} |\widehat{f}(\xi, \eta)|^2 \, d\xi \, d\eta.$$

Finally,

$$(A.23.9) \quad \int_{|\xi| > R, \, |\eta| > S} |\eta|^{2\mu} |\widehat{f}(\xi, \eta)|^2 \, d\xi \, d\eta \leq \frac{1}{S^{2(\mu' - \mu)}} \int |\eta|^{2\mu'} |\widehat{f}(\xi, \eta)|^2 \, d\xi \, d\eta.$$

Choose R and S such that $S^{2\mu}/R^{2\lambda'} = 1/S^{2(\mu' - \mu)}$, i.e. $R = S^{\mu'/\lambda'}$. This yields a bound like

$$C_n S^{2\mu + n \left(1 + \frac{\mu'}{\lambda'}\right)} \|\widehat{f}\|_{L^\infty}^2 + S^{2(\mu - \mu')} \left(\int |\xi|^{2\lambda'} |\widehat{f}|^2 + \int |\eta|^{2\mu'} |\widehat{f}|^2 \right).$$

Then the result follows by optimization in S.

By symmetry, the same argument works for the case when $\mu = 0$. Now for the general case, we first choose p and q such that $p^{-1} + q^{-1} = 1$ and $p^{-1} \geq \lambda/\lambda'$, $q^{-1} \geq \mu/\mu'$, and apply Hölder's inequality with conjugate exponents p and q:

$$(A.23.10) \quad \int |\xi|^{2\lambda} |\eta|^{2\mu} |\widehat{f}|^2 \, d\xi \, d\eta \leq \left(\int |\xi|^{2\lambda p} |\widehat{f}|^2 \, d\xi \, d\eta \right)^{\frac{1}{p}} \left(\int |\eta|^{2\mu q} |\widehat{f}|^2 \, d\xi \, d\eta \right)^{\frac{1}{q}}.$$

Then we apply to the integrals in the right-hand side of (A.23.10) the results obtained before for $\lambda = 0$ and $\mu = 0$:

$$\int |\xi|^{2\lambda p} |\widehat{f}|^2 \leq C \left(\int |\xi|^{2\lambda'} |\widehat{f}|^2 + \int |\eta|^{2\mu'} |\widehat{f}|^2 \right)^{1 - \theta_1} \|\widehat{f}\|_{L^\infty}^{2\theta_1},$$

and
$$\int |\eta|^{2\mu q}|\widehat{f}|^2 \leq C \left(\int |\xi|^{2\lambda'}|\widehat{f}|^2 + \int |\eta|^{2\mu'}|\widehat{f}|^2\right)^{1-\theta_2} \|\widehat{f}\|_{L^\infty}^{2\theta_2},$$
where
$$\theta_1 = \frac{\lambda' - \lambda p}{\lambda + \frac{n}{2}\left(1 + \frac{\lambda'}{\mu'}\right)}, \qquad \theta_2 = \frac{\mu' - \mu q}{\mu + \frac{n}{2}\left(1 + \frac{\mu'}{\lambda'}\right)}.$$

After some calculation, one finds
$$\frac{\theta_1}{p} + \frac{\theta_2}{q} = \frac{1 - \left(\frac{\lambda}{\lambda'} + \frac{\mu}{\mu'}\right)}{1 + \frac{n}{2}\left(\frac{1}{\lambda'} + \frac{1}{\mu'}\right)},$$
and the result follows. □

The next technical lemma in this Appendix is an estimate about a system of differential inequalities. The system may look rather peculiar, but I believe that it arises naturally in many problems of hypoelliptic regularization. In any case, it is used in various places of Appendix A.21.

LEMMA A.26. *Let $\mathcal{E}$, X, Y, Z and $\mathcal{M}$ be continuous functions of $t \in [0, 1]$, with $\mathcal{E}, X, Y, Z \geq 0$, such that*

(A.23.11) $$K(X + Y) \leq \mathcal{E} \leq C(X + Y),$$

(A.23.12) $$|\mathcal{M}| \leq C\mathcal{E}^{1-\delta},$$

(A.23.13) $$\frac{d\mathcal{E}}{dt} \leq -KZ + C\mathcal{E},$$

(A.23.14) $$Y \leq C(X + Z)^{1-\theta},$$

(A.23.15) $$\frac{d\mathcal{M}}{dt} \leq -KX + C(Y + Z),$$

where C, K are positive constants, and δ, θ are real numbers lying in $(0, 1)$. Then
$$\mathcal{E}(t) \leq \frac{\overline{C}}{t^{1/\kappa}}, \qquad \kappa = \min\left(\delta, \frac{\theta}{1-\theta}\right),$$
where $\overline{C}$ is an explicit constant which only depends on C, K, θ, δ.

PROOF OF LEMMA A.26. Let $\widetilde{\mathcal{E}}(t) = e^{-Ct}\mathcal{E}(t)$; then $\widetilde{\mathcal{E}}$ satisfies estimates similar to $\mathcal{E}$, except that equation (A.23.13) becomes $d\widetilde{\mathcal{E}}/dt \leq -KZ$. In the sequel I shall keep the notation $\mathcal{E}$ for $\widetilde{\mathcal{E}}$, so this just amounts to replacing (A.23.13) by

(A.23.16) $$\frac{d\mathcal{E}}{dt} \leq -KZ.$$

In particular, $\mathcal{E}$ is nonincreasing.

Now let $E > 0$, and let $I \subset [0, 1]$ be the time-interval where $(E/2) \leq \mathcal{E}(t) \leq E$. The goal is to show that the length $|I|$ of I is bounded like $O(E^{-\kappa})$ for some $\kappa > 0$. If that is the case, then the conclusion follows. Indeed, let $E_0 > 0$ be given, and let T be the first time t such that $\mathcal{E}(t) \leq E_0$, then
$$T \leq C' \sum_{n \geq 1} E_0^{-n\kappa} \leq C'' E_0^{-\kappa};$$

so $E_0 \leq T^{-1/\kappa}$. (Here as in the sequel, C, C', C'' stand for various constants that only depend on the constants C and K appearing in the statement of the lemma.)

If $E \leq 1$ then the conclusion obviously holds true. So we might assume that $E \geq 1$.

It follows by integration of (A.23.16) over I that

(A.23.17) $$\int_I Z(t)\, dt \leq E - \frac{E}{2} = \frac{E}{2}.$$

By integrating (A.23.14), we find

$$\int_I Y(t)\, dt \leq C \int_I [X(t) + Z(t)]^{1-\theta}\, dt$$
$$\leq C' \left(\int_I X(t)^{1-\theta}\, dt + \int_I Z(t)^{1-\theta}\, dt \right)$$
$$\leq C' \left(|I| \left[\sup_I X(t)^{1-\theta} \right] + \left(\int_I Z(t)\, dt \right)^{1-\theta} |I|^\theta \right).$$

To estimate the first term inside parentheses, note that $X \leq C\mathcal{E} \leq CE$; to bound the second term, use (A.23.17). The result is

(A.23.18) $$\int_I Y(t)\, dt \leq C\left(|I| E^{1-\theta} + E^{1-\theta} |I|^\theta \right) \leq C' |I|^\theta E^{1-\theta},$$

where the last inequality follows from $|I| \leq |I|^\theta$. (Note indeed that $|I| \leq 1$ and $\theta < 1$.)

Next, integrate inequality (A.23.15) over $I = [t_1, t_2]$, to get

(A.23.19) $$K \int_I X(t)\, dt \leq |\mathcal{M}(t_1)| + |\mathcal{M}(t_2)| + C \int_I [Y(t) + Z(t)]\, dt$$

(A.23.20) $$\leq 2 \sup_{t \in I} |\mathcal{M}(t)| + C \left(\int_I Y(t)\, dt + \int_I Z(t)\, dt \right).$$

Also, since $\mathcal{E} \geq E/2$ on I, we have

(A.23.21) $$\frac{|I| E}{2} \leq \int_I \mathcal{E}(t)\, dt \leq C \left(\int_I X(t)\, dt + \int_I Y(t)\, dt \right),$$

where the last inequality follows from (A.23.11).

The combination of (A.23.20) and (A.23.21) implies

$$\frac{|I| E}{2} \leq C \left(\sup_{t \in I} |\mathcal{M}(t)| + \int_I Y(t)\, dt + \int_I Z(t)\, dt \right).$$

To estimate the first term inside the brackets, use (A.23.14); to estimate the second one, use (A.23.18); to estimate the third one, use (A.23.17). The result is

(A.23.22) $$|I| E \leq C(E^{1-\delta} + |I|^\theta E^{1-\theta} + E).$$

Now we can conclude, separating three cases according to which one of the three terms in the right-hand side of (A.23.22) is largest:

- If it is $E^{1-\delta}$, then $|I| E \leq 3C E^{1-\delta}$, so $|I| \leq 3C E^{-\delta}$;
- If it is $|I|^\theta E^{1-\theta}$, then $|I| E \leq 3C |I|^\theta E^{1-\theta}$, so $|I| \leq (3C)^{\frac{1}{1-\theta}} E^{-\frac{\theta}{1-\theta}}$;
- If it is E, then $|I| \leq 3C$.

In any case, there is an estimate like $|I| \leq \overline{C} E^{-\kappa}$, where κ is as in the statement of the lemma. So the proof is complete. □

The final result in this appendix is a variation of the usual Korn inequality, used in Subsection 18.5.

PROPOSITION A.27 (trace Korn inequality). *Let Ω be a smooth bounded connected open subset of $\mathbb{R}^N$. Then there is a constant $C = C(\Omega)$ such that for any vector field $u \in H^1(\Omega; \mathbb{R}^N)$, tangent to the boundary $\partial\Omega$,*

(A.23.23) $$\|\nabla u\|_{L^2(\Omega)}^2 \leq C\bigl(\|\nabla^{\mathrm{sym}} u\|_{L^2(\Omega)} + \|u\|_{L^2(\partial\Omega)}\bigr),$$

where $\nabla^{\mathrm{sym}} u$ stands for the symmetric part of the matrix-valued field ∇u.

PROOF. By density, we may assume that u is smooth. According to [**15**, eq. (39)-(42)], if u is tangent to the boundary, then

$$\int_\Omega |\nabla^{\mathrm{sym}} u|^2 = \int_\Omega |\nabla^{\mathrm{a}} u|^2 + \int_\Omega (\nabla \cdot u)^2 - \int_{\partial\Omega} (\mathrm{II})_\Omega(u,u),$$

where $\nabla^{\mathrm{a}} u$ stands for the antisymmetric part of ∇u, and $(\mathrm{II})_\Omega$ for the second fundamental form of the domain Ω. It follows that

$$\int_\Omega |\nabla^{\mathrm{a}} u|^2 \leq \int_\Omega |\nabla^{\mathrm{sym}} u|^2 + C \int_{\partial\Omega} |u|^2,$$

where $C = \max_{\partial\Omega} \|(\mathrm{II})_\Omega\|$. Inequality (A.23.23) follows immediately. □

Bibliography

[1] ARKERYD, L. The stationary Boltzmann equation with diffuse reflection boundary values. *İstanbul Tek. Üniv. Bül.* **47**, 1-2 (1994), 209–217. MR1321952 (96c:82048)

[2] ARKERYD, L., AND NOURI, A. The stationary Boltzmann equation in $\mathbb{R}^n$ with given indata. *Ann. Sc. Norm. Super. Pisa Cl. Sci. (5)* **1**, 2 (2002), 359–385. MR1991144 (2004m:35040)

[3] ARKERYD, L., AND NOURI, A. A large data existence result for the stationary Boltzmann equation in a cylindrical geometry. *Ark. Mat.* **43**, 1 (2005), 29–50. MR2134697 (2005m:35034)

[4] ARNOLD, A., MARKOWICH, P., TOSCANI, G., AND UNTERREITER, A. On logarithmic Sobolev inequalities and the rate of convergence to equilibrium for Fokker–Planck type equations. *Comm. Partial Differential Equations* **26**, 1–2 (2001), 43–100. MR1842428 (2002d:35097)

[5] ARNOL′D, V. *Équations différentielles ordinaires*, in French, fourth ed. "Mir", Moscow, 1988. MR990889 (91b:34001)

[6] BOBKOV, S. G., AND LEDOUX, M. From Brunn–Minkowski to Brascamp–Lieb and to logarithmic Sobolev inequalities. *Geom. Funct. Anal.* **10**, 5 (2000), 1028–1052. MR1800062 (2002k:26028)

[7] BOUCHUT, F. Hypoelliptic regularity in kinetic equations. *J. Math. Pures Appl. (9)* **81**, 11 (2002), 1135–1159. MR1949176 (2003k:82081)

[8] BRASCAMP, H. J., AND LIEB, E. H. On extensions of the Brunn–Minkowski and Prékopa–Leindler theorems, including inequalities for log concave functions, and with an application to the diffusion equation. *J. Functional Analysis* **22**, 4 (1976), 366–389. MR0450480 (56:8774)

[9] CÁCERES, M. J., CARRILLO, J. A., AND GOUDON, T. Equilibration rate for the linear inhomogeneous relaxation-time Boltzmann equation for charged particles. *Comm. Partial Differential Equations* **28**, 5-6 (2003), 969–989. MR1986057 (2004g:82111)

[10] CAPELLA, A., LOESCHCKE, C., AND WACHSMUTH, J. On the dissipation of the linearized LLG-Maxwell equations with eddy current damping. Draft note, 2006.

[11] CERCIGNANI, C. *Rarefied gas dynamics. From basic concepts to actual calculations.* Cambridge University Press, Cambridge, 2000. MR1744523 (2001f:76065)

[12] DESVILLETTES, L. Convergence to equilibrium in large time for Boltzmann and BGK equations. *Arch. Rational Mech. Anal.* **110**, 1 (1990), 73–91. MR1031086 (91d:35210)

[13] DESVILLETTES, L., AND VILLANI, C. On the spatially homogeneous Landau equation for hard potentials. I. Existence, uniqueness and smoothness. *Comm. Partial Differential Equations* **25**, 1-2 (2000), 179–259. MR1737547 (2001c:82065)

[14] DESVILLETTES, L., AND VILLANI, C. On the trend to global equilibrium in spatially inhomogeneous entropy-dissipating systems: the linear Fokker–Planck equation. *Comm. Pure Appl. Math.* **54**, 1 (2001), 1–42. MR1787105 (2001h:82079)

[15] DESVILLETTES, L., AND VILLANI, C. On a variant of Korn's inequality arising in statistical mechanics. *ESAIM Control Optim. Calc. Var.* **8** (2002), 603–619. MR1932965 (2004i:82053)

[16] DESVILLETTES, L., AND VILLANI, C. On the trend to global equilibrium for spatially inhomogeneous kinetic systems: the Boltzmann equation. *Invent. Math.* **159**, 2 (2005), 245–316. MR2116276 (2005j:82070)

[17] DEUSCHEL, J.-D., AND STROOCK, D. W. *Large deviations*, vol. 137 of *Pure and Applied Mathematics*. Academic Press Inc., Boston, MA, 1989. MR997938 (90h:60026)

[18] E, W., MATTINGLY, J. C., AND SINAI, Y. Gibbsian dynamics and ergodicity for the stochastically forced Navier–Stokes equation. *Comm. Math. Phys.* **224**, 1 (2001), 83–106. MR1868992 (2002m:76024)

[19] ECKMANN, J.-P., AND HAIRER, M. Uniqueness of the invariant measure for a stochastic PDE driven by degenerate noise. *Comm. Math. Phys.* **219**, 3 (2001), 523–565. MR1838749 (2002d:60054)

[20] ECKMANN, J.-P., AND HAIRER, M. Spectral properties of hypoelliptic operators. *Comm. Math. Phys. 235*, 2 (2003), 233–253. MR1969727 (2004c:35060)

[21] ECKMANN, J.-P., PILLET, C.-A., AND REY-BELLET, L. Non-equilibrium statistical mechanics of anharmonic chains coupled to two heat baths at different temperatures. *Comm. Math. Phys. 201*, 3 (1999), 657–697. MR1685893 (2000d:82025)

[22] FELLNER, K., NEUMANN, L., AND SCHMEISER, C. Convergence to global equilibrium for spatially inhomogeneous kinetic models of non-micro-reversible processes. *Monatsh. Math. 141*, 4 (2004), 289–299. MR2053654 (2005f:82116)

[23] FILBET, F., MOUHOT, C., AND PARESCHI, L. Solving the Boltzmann equation in $N \log_2 N$. *SIAM J. Sci. Comput. 28*, 3 (2006), 1029–1053. MR2240802 (2007c:82076)

[24] GALLAGHER, I., GALLAY, T., AND NIER, F. Spectral asymptotics for large skew-symmetric perturbations of the harmonic oscillator. Preprint, 2008.

[25] GALLAY, T., AND WAYNE, C. E. Invariant manifolds and the long-time asymptotics of the Navier–Stokes and vorticity equations on $\mathbf{R}^2$. *Arch. Ration. Mech. Anal. 163*, 3 (2002), 209–258. MR1912106 (2003c:37123)

[26] GALLAY, T., AND WAYNE, C. E. Global stability of vortex solutions of the two-dimensional Navier–Stokes equation. *Comm. Math. Phys. 255*, 1 (2005), 97–129. MR2123378 (2005m:35224)

[27] GRIGOR'YAN, A. A. Analytic and geometric background of recurrence and non-explosion of the Brownian motion on Riemannian manifolds. *Bull. Amer. Math. Soc. 36*, 2 (1999), 135–249. MR1659871 (99k:58195)

[28] GUO, Y. The Landau equation in a periodic box. *Comm. Math. Phys. 231*, 3 (2002), 391–434. MR1946444 (2004c:82121)

[29] GUO, Y., AND STRAIN, R. M. Exponential decay for soft potentials near Maxwellian. *Arch. Ration. Mech. Anal. 187*, 2 (2008), 287–339. MR2366140 (2008m:82008)

[30] HAIRER, M., AND MATTINGLY, J. C. Ergodicity of the 2D Navier–Stokes equations with degenerate stochastic forcing. *Ann. of Maths (2) 164*, 3 (2006), 993–1032. MR2259251 (2008a:37095)

[31] HANOUZET, B., AND NATALINI, R. Global existence of smooth solutions for partially dissipative hyperbolic systems with a convex entropy. *Arch. Ration. Mech. Anal. 169*, 2 (2003), 89–117. MR2005637 (2004h:35135)

[32] HELFFER, B., AND NIER, F. In *Hypoellipticity and spectral theory for Fokker–Planck operators and Witten Laplacians*, vol. 1862 of *Lecture Notes in Math.* Springer, Berlin, 2005. MR2130405 (2006a:58039)

[33] HÉRAU, F. Short and long time behavior of the Fokker–Planck equation in a confining potential and applications. *J. Funct. Anal. 244*, 1 (2007), 95–118. MR2294477 (2008e:47099)

[34] HÉRAU, F., AND NIER, F. Isotropic hypoellipticity and trend to equilibrium for the Fokker–Planck equation with a high-degree potential. *Arch. Ration. Mech. Anal. 171*, 2 (2004), 151–218. MR2034753 (2005f:82085)

[35] HÖRMANDER, L. Hypoelliptic second order differential equations. *Acta Math. 119* (1967), 147–171. MR0222474 (36:5526)

[36] HÖRMANDER, L. Hypoelliptic second order differential equations. *Acta Math. 119* (1967), 147–171. MR0222474 (36:5526)

[37] KAWASHIMA, S. Large-time behaviour of solutions to hyperbolic-parabolic systems of conservation laws and applications. *Proc. Roy. Soc. Edinburgh Sect. A 106*, 1-2 (1987), 169–194. MR899951 (89d:35022)

[38] KOHN, J. Pseudo-differential operators and hypoellipticity. In *Proc. Symp. Pure Math.* (1969), vol. 23, AMS Providence, RI, pp. 61–69. MR0338592 (49:3356)

[39] KOLMOGOROV, A. Zufällige Bewgungen (zur Theorie der Brownschen Bewegung). *Ann. of Math. (2) 35*, 1 (1934), 116–117. MR1503147

[40] MATTINGLY, J. C. Exponential convergence for the stochastically forced Navier–Stokes equations and other partially dissipative dynamics. *Comm. Math. Phys. 230*, 3 (2002), 421–462. MR1937652 (2004a:76039)

[41] MATTINGLY, J. C., STUART, A. M., AND HIGHAM, D. J. Ergodicity for SDEs and approximations: locally Lipschitz vector fields and degenerate noise. *Stochastic Process. Appl. 101*, 2 (2002), 185–232. MR1931266 (2003i:60103)

[42] MOUHOT, C. Quantitative lower bounds for the full Boltzmann equation. I. Periodic boundary conditions. *Comm. Partial Differential Equations 30*, 4-6 (2005), 881–917. MR2153518 (2006a:76096)

[43] MOUHOT, C., AND NEUMANN, L. Quantitative perturbative study of convergence to equilibrium for collisional kinetic models in the torus. *Nonlinearity 19*, 4 (2006), 969–998. MR2214953 (2007c:82032)

[44] NASH, J. Continuity of solutions of parabolic and elliptic equations. *Amer. J. Math. 80* (1958), 931–954. MR0100158 (20:6592)

[45] REY-BELLET, L., AND THOMAS, L. E. Asymptotic behavior of thermal nonequilibrium steady states for a driven chain of anharmonic oscillators. *Comm. Math. Phys. 215*, 1 (2000), 1–24. MR1799873 (2001k:82061)

[46] REY-BELLET, L., AND THOMAS, L. E. Exponential convergence to non-equilibrium stationary states in classical statistical mechanics. *Comm. Math. Phys. 225*, 2 (2002), 305–329. MR1889227 (2003f:82052)

[47] RISKEN, H. *The Fokker–Planck equation*, second ed., vol. 18 of *Springer Series in Synergetics*. Springer-Verlag, Berlin, 1989. MR987631 (90a:82002)

[48] ROCKNER, M., AND WANG, F.-Y. On the spectrum of a class of non-sectorial diffusion operators. *Bull. London Math. Soc. 36* (2004), 95–104. MR2011983 (2004j:31013)

[49] ROTHSCHILD, L. P., AND STEIN, E. M. Hypoelliptic differential operators and nilpotent groups. *Acta Math. 137*, 3-4 (1976), 247–320. MR0436223 (55:9171)

[50] RUGGERI, T., AND SERRE, D. Stability of constant equilibrium state for dissipative balance laws system with a convex entropy. *Quart. Appl. Math. 62*, 1 (2004), 163–179. MR2032577 (2004k:35257)

[51] SALOFF-COSTE, L. *Aspects of Sobolev-type inequalities*, vol. 289 of *London Mathematical Society Lecture Note Series*. Cambridge University Press, Cambridge, 2002. MR1872526 (2003c:46048)

[52] STRAIN, R. M., AND GUO, Y. Almost exponential decay near Maxwellian. *Comm. Partial Differential Equations 31*, 1-3 (2006), 417–429. MR2209761 (2006m:82042)

[53] TALAY, D. Stochastic Hamiltonian systems: exponential convergence to the invariant measure, and discretization by the implicit Euler scheme. In *Inhomogeneous random systems (Cergy-Pontoise, 2001)*. Markov Process. Related Fields 8, 2 (2002), 163–198. MR1924934 (2003e:60129)

[54] TOSCANI, G., AND VILLANI, C. Sharp entropy dissipation bounds and explicit rate of trend to equilibrium for the spatially homogeneous Boltzmann equation. *Comm. Math. Phys. 203*, 3 (1999), 667–706. MR1700142 (2000e:82039)

[55] TOSCANI, G., AND VILLANI, C. On the trend to equilibrium for some dissipative systems with slowly increasing a priori bounds. *J. Statist. Phys. 98*, 5-6 (2000), 1279–1309. MR1751701 (2001g:82069)

[56] VILLANI, C. Entropy dissipation and convergence to equilibrium. Notes from a series of lectures in Institut Henri Poincaré, Paris (2001). Updated in 2007. *Lecture Notes in Mathematics*, Vol. 1916. Springer, Berlin, 2008. MR2409050

[57] VILLANI, C. A review of mathematical topics in collisional kinetic theory. In *Handbook of mathematical fluid dynamics, Vol. I*. North-Holland, Amsterdam, 2002, pp. 71–305. MR1942465 (2003k:82087)

[58] VILLANI, C. Cercignani's conjecture is sometimes true and always almost true. *Comm. Math. Phys. 234*, 3 (2003), 455–490. MR1964379 (2004b:82048)

Editorial Information

To be published in the *Memoirs*, a paper must be correct, new, nontrivial, and significant. Further, it must be well written and of interest to a substantial number of mathematicians. Piecemeal results, such as an inconclusive step toward an unproved major theorem or a minor variation on a known result, are in general not acceptable for publication.

Papers appearing in *Memoirs* are generally at least 80 and not more than 200 published pages in length. Papers less than 80 or more than 200 published pages require the approval of the Managing Editor of the Transactions/Memoirs Editorial Board. Published pages are the same size as those generated in the style files provided for $\mathcal{AMS}$-L^{A}T$_E$X or $\mathcal{AMS}$-T$_E$X.

Information on the backlog for this journal can be found on the AMS website starting from http://www.ams.org/memo.

A Consent to Publish and Copyright Agreement is required before a paper will be published in the *Memoirs*. After a paper is accepted for publication, the Providence office will send a Consent to Publish and Copyright Agreement to all authors of the paper. By submitting a paper to the *Memoirs*, authors certify that the results have not been submitted to nor are they under consideration for publication by another journal, conference proceedings, or similar publication.

Information for Authors

Memoirs is an author-prepared publication. Once formatted for print and on-line publication, articles will be published as is with the addition of AMS-prepared frontmatter and backmatter. Articles are not copyedited; however, confirmation copy will be sent to the authors.

Initial submission. The AMS uses Centralized Manuscript Processing for initial submissions. Authors should submit a PDF file using the Initial Manuscript Submission form found at www.ams.org/peer-review-submission, or send one copy of the manuscript to the following address: Centralized Manuscript Processing, MEMOIRS OF THE AMS, 201 Charles Street, Providence, RI 02904-2294 USA. If a paper copy is being forwarded to the AMS, indicate that it is for *Memoirs* and include the name of the corresponding author, contact information such as email address or mailing address, and the name of an appropriate Editor to review the paper (see the list of Editors below).

The paper must contain a *descriptive title* and an *abstract* that summarizes the article in language suitable for workers in the general field (algebra, analysis, etc.). The *descriptive title* should be short, but informative; useless or vague phrases such as "some remarks about" or "concerning" should be avoided. The *abstract* should be at least one complete sentence, and at most 300 words. Included with the footnotes to the paper should be the 2010 *Mathematics Subject Classification* representing the primary and secondary subjects of the article. The classifications are accessible from www.ams.org/msc/. The Mathematics Subject Classification footnote may be followed by a list of *key words and phrases* describing the subject matter of the article and taken from it. Journal abbreviations used in bibliographies are listed in the latest *Mathematical Reviews* annual index. The series abbreviations are also accessible from www.ams.org/msnhtml/serials.pdf. To help in preparing and verifying references, the AMS offers MR Lookup, a Reference Tool for Linking, at www.ams.org/mrlookup/.

Electronically prepared manuscripts. The AMS encourages electronically prepared manuscripts, with a strong preference for $\mathcal{AMS}$-L^{A}T$_E$X. To this end, the Society has prepared $\mathcal{AMS}$-L^{A}T$_E$X author packages for each AMS publication. Author packages include instructions for preparing electronic manuscripts, samples, and a style file that generates the particular design specifications of that publication series. Though $\mathcal{AMS}$-L^{A}T$_E$X is the highly preferred format of T$_E$X, author packages are also available in $\mathcal{AMS}$-T$_E$X.

Authors may retrieve an author package for *Memoirs of the AMS* from www.ams.org/journals/memo/memoauthorpac.html or via FTP to ftp.ams.org (login as anonymous, enter your complete email address as password, and type cd pub/author-info). The

AMS Author Handbook and the *Instruction Manual* are available in PDF format from the author package link. The author package can also be obtained free of charge by sending email to tech-support@ams.org (Internet) or from the Publication Division, American Mathematical Society, 201 Charles St., Providence, RI 02904-2294, USA. When requesting an author package, please specify $\mathcal{AMS}$-LaTeX or $\mathcal{AMS}$-TeX and the publication in which your paper will appear. Please be sure to include your complete mailing address.

After acceptance. The source files for the final version of the electronic manuscript should be sent to the Providence office immediately after the paper has been accepted for publication. The author should also submit a PDF of the final version of the paper to the editor, who will forward a copy to the Providence office.

Accepted electronically prepared files can be submitted via the web at www.ams.org/submit-book-journal/, sent via FTP, or sent on CD-Rom or diskette to the Electronic Prepress Department, American Mathematical Society, 201 Charles Street, Providence, RI 02904-2294 USA. TeX source files and graphic files can be transferred over the Internet by FTP to the Internet node ftp.ams.org (130.44.1.100). When sending a manuscript electronically via CD-Rom or diskette, please be sure to include a message indicating that the paper is for the *Memoirs*.

Electronic graphics. Comprehensive instructions on preparing graphics are available at www.ams.org/authors/journals.html. A few of the major requirements are given here.

Submit files for graphics as EPS (Encapsulated PostScript) files. This includes graphics originated via a graphics application as well as scanned photographs or other computer-generated images. If this is not possible, TIFF files are acceptable as long as they can be opened in Adobe Photoshop or Illustrator.

Authors using graphics packages for the creation of electronic art should also avoid the use of any lines thinner than 0.5 points in width. Many graphics packages allow the user to specify a "hairline" for a very thin line. Hairlines often look acceptable when proofed on a typical laser printer. However, when produced on a high-resolution laser imagesetter, hairlines become nearly invisible and will be lost entirely in the final printing process.

Screens should be set to values between 15% and 85%. Screens which fall outside of this range are too light or too dark to print correctly. Variations of screens within a graphic should be no less than 10%.

Inquiries. Any inquiries concerning a paper that has been accepted for publication should be sent to memo-query@ams.org or directly to the Electronic Prepress Department, American Mathematical Society, 201 Charles St., Providence, RI 02904-2294 USA.

Editors

This journal is designed particularly for long research papers, normally at least 80 pages in length, and groups of cognate papers in pure and applied mathematics. Papers intended for publication in the *Memoirs* should be addressed to one of the following editors. The AMS uses Centralized Manuscript Processing for initial submissions to AMS journals. Authors should follow instructions listed on the Initial Submission page found at www.ams.org/memo/memosubmit.html.

Algebra, to ALEXANDER KLESHCHEV, Department of Mathematics, University of Oregon, Eugene, OR 97403-1222; e-mail: ams@noether.uoregon.edu

Algebraic geometry, to DAN ABRAMOVICH, Department of Mathematics, Brown University, Box 1917, Providence, RI 02912; e-mail: amsedit@math.brown.edu

Algebraic geometry and its applications, to MINA TEICHER, Emmy Noether Research Institute for Mathematics, Bar-Ilan University, Ramat-Gan 52900, Israel; e-mail: teicher@macs.biu.ac.il

Algebraic topology, to ALEJANDRO ADEM, Department of Mathematics, University of British Columbia, Room 121, 1984 Mathematics Road, Vancouver, British Columbia, Canada V6T 1Z2; e-mail: adem@math.ubc.ca

Combinatorics, to JOHN R. STEMBRIDGE, Department of Mathematics, University of Michigan, Ann Arbor, Michigan 48109-1109; e-mail: JRS@umich.edu

Commutative and homological algebra, to LUCHEZAR L. AVRAMOV, Department of Mathematics, University of Nebraska, Lincoln, NE 68588-0130; e-mail: avramov@math.unl.edu

Complex analysis and harmonic analysis, to ALEXANDER NAGEL, Department of Mathematics, University of Wisconsin, 480 Lincoln Drive, Madison, WI 53706-1313; e-mail: nagel@math.wisc.edu

Differential geometry and global analysis, to CHRIS WOODWARD, Department of Mathematics, Rutgers University, 110 Frelinghuysen Road, Piscataway, NJ 08854; e-mail: ctw@math.rutgers.edu

Dynamical systems and ergodic theory and complex analysis, to YUNPING JIANG, Department of Mathematics, CUNY Queens College and Graduate Center, 65-30 Kissena Blvd., Flushing, NY 11367; e-mail: Yunping.Jiang@qc.cuny.edu

Functional analysis and operator algebras, to DIMITRI SHLYAKHTENKO, Department of Mathematics, University of California, Los Angeles, CA 90095; e-mail: shlyakht@math.ucla.edu

Geometric analysis, to WILLIAM P. MINICOZZI II, Department of Mathematics, Johns Hopkins University, 3400 N. Charles St., Baltimore, MD 21218; e-mail: trans@math.jhu.edu

Geometric topology, to MARK FEIGHN, Math Department, Rutgers University, Newark, NJ 07102; e-mail: feighn@andromeda.rutgers.edu

Harmonic analysis, representation theory, and Lie theory, to ROBERT J. STANTON, Department of Mathematics, The Ohio State University, 231 West 18th Avenue, Columbus, OH 43210-1174; e-mail: stanton@math.ohio-state.edu

Logic, to STEFFEN LEMPP, Department of Mathematics, University of Wisconsin, 480 Lincoln Drive, Madison, Wisconsin 53706-1388; e-mail: lempp@math.wisc.edu

Number theory, to JONATHAN ROGAWSKI, Department of Mathematics, University of California, Los Angeles, CA 90095; e-mail: jonr@math.ucla.edu

Number theory, to SHANKAR SEN, Department of Mathematics, 505 Malott Hall, Cornell University, Ithaca, NY 14853; e-mail: ss70@cornell.edu

Partial differential equations, to GUSTAVO PONCE, Department of Mathematics, South Hall, Room 6607, University of California, Santa Barbara, CA 93106; e-mail: ponce@math.ucsb.edu

Partial differential equations and dynamical systems, to PETER POLACIK, School of Mathematics, University of Minnesota, Minneapolis, MN 55455; e-mail: polacik@math.umn.edu

Probability and statistics, to RICHARD BASS, Department of Mathematics, University of Connecticut, Storrs, CT 06269-3009; e-mail: bass@math.uconn.edu

Real analysis and partial differential equations, to DANIEL TATARU, Department of Mathematics, University of California, Berkeley, Berkeley, CA 94720; e-mail: tataru@math.berkeley.edu

All other communications to the editors, should be addressed to the Managing Editor, ROBERT GURALNICK, Department of Mathematics, University of Southern California, Los Angeles, CA 90089-1113; e-mail: guralnic@math.usc.edu.

Titles in This Series

951 **Pierre Magal and Shigui Ruan,** Center manifolds for semilinear equations with non-dense domain and applications to Hopf bifurcation in age structured models, 2009

950 **Cédric Villani,** Hypocoercivity, 2009

949 **Drew Armstrong,** Generalized noncrossing partitions and combinatorics of Coxeter groups, 2009

948 **Nan-Kuo Ho and Chiu-Chu Melissa Liu,** Yang-Mills connections on orientable and nonorientable surfaces, 2009

947 **W. Turner,** Rock blocks, 2009

946 **Jay Jorgenson and Serge Lang,** Heat Eisenstein series on $SL_n(C)$, 2009

945 **Tobias H. Jäger,** The creation of strange non-chaotic attractors in non-smooth saddle-node bifurcations, 2009

944 **Yuri Kifer,** Large deviations and adiabatic transitions for dynamical systems and Markov processes in fully coupled averaging, 2009

943 **István Berkes and Michel Weber,** On the convergence of $\sum c_k f(n_k x)$, 2009

942 **Dirk Kussin,** Noncommutative curves of genus zero: Related to finite dimensional algebras, 2009

941 **Gelu Popescu,** Unitary invariants in multivariable operator theory, 2009

940 **Gérard Iooss and Pavel I. Plotnikov,** Small divisor problem in the theory of three-dimensional water gravity waves, 2009

939 **I. D. Suprunenko,** The minimal polynomials of unipotent elements in irreducible representations of the classical groups in odd characteristic, 2009

938 **Antonino Morassi and Edi Rosset,** Uniqueness and stability in determining a rigid inclusion in an elastic body, 2009

937 **Skip Garibaldi,** Cohomological invariants: Exceptional groups and spin groups, 2009

936 **André Martinez and Vania Sordoni,** Twisted pseudodifferential calculus and application to the quantum evolution of molecules, 2009

935 **Mihai Ciucu,** The scaling limit of the correlation of holes on the triangular lattice with periodic boundary conditions, 2009

934 **Arjen Doelman, Björn Sandstede, Arnd Scheel, and Guido Schneider,** The dynamics of modulated wave trains, 2009

933 **Luchezar Stoyanov,** Scattering resonances for several small convex bodies and the Lax-Phillips conjuecture, 2009

932 **Jun Kigami,** Volume doubling measures and heat kernel estimates of self-similar sets, 2009

931 **Robert C. Dalang and Marta Sanz-Solé,** Hölder-Sobolv regularity of the solution to the stochastic wave equation in dimension three, 2009

930 **Volkmar Liebscher,** Random sets and invariants for (type II) continuous tensor product systems of Hilbert spaces, 2009

929 **Richard F. Bass, Xia Chen, and Jay Rosen,** Moderate deviations for the range of planar random walks, 2009

928 **Ulrich Bunke,** Index theory, eta forms, and Deligne cohomology, 2009

927 **N. Chernov and D. Dolgopyat,** Brownian Brownian motion-I, 2009

926 **Riccardo Benedetti and Francesco Bonsante,** Canonical wick rotations in 3-dimensional gravity, 2009

925 **Sergey Zelik and Alexander Mielke,** Multi-pulse evolution and space-time chaos in dissipative systems, 2009

924 **Pierre-Emmanuel Caprace,** "Abstract" homomorphisms of split Kac-Moody groups, 2009

923 **Michael Jöllenbeck and Volkmar Welker,** Minimal resolutions via algebraic discrete Morse theory, 2009

922 **Ph. Barbe and W. P. McCormick,** Asymptotic expansions for infinite weighted convolutions of heavy tail distributions and applications, 2009

TITLES IN THIS SERIES

- 921 **Thomas Lehmkuhl,** Compactification of the Drinfeld modular surfaces, 2009
- 920 **Georgia Benkart, Thomas Gregory, and Alexander Premet,** The recognition theorem for graded Lie algebras in prime characteristic, 2009
- 919 **Roelof W. Bruggeman and Roberto J. Miatello,** Sum formula for SL_2 over a totally real number field, 2009
- 918 **Jonathan Brundan and Alexander Kleshchev,** Representations of shifted Yangians and finite W-algebras, 2008
- 917 **Salah-Eldin A. Mohammed, Tusheng Zhang, and Huaizhong Zhao,** The stable manifold theorem for semilinear stochastic evolution equations and stochastic partial differential equations, 2008
- 916 **Yoshikata Kida,** The mapping class group from the viewpoint of measure equivalence theory, 2008
- 915 **Sergiu Aizicovici, Nikolaos S. Papageorgiou, and Vasile Staicu,** Degree theory for operators of monotone type and nonlinear elliptic equations with inequality constraints, 2008
- 914 **E. Shargorodsky and J. F. Toland,** Bernoulli free-boundary problems, 2008
- 913 **Ethan Akin, Joseph Auslander, and Eli Glasner,** The topological dynamics of Ellis actions, 2008
- 912 **Igor Chueshov and Irena Lasiecka,** Long-time behavior of second order evolution equations with nonlinear damping, 2008
- 911 **John Locker,** Eigenvalues and completeness for regular and simply irregular two-point differential operators, 2008
- 910 **Joel Friedman,** A proof of Alon's second eigenvalue conjecture and related problems, 2008
- 909 **Cameron McA. Gordon and Ying-Qing Wu,** Toroidal Dehn fillings on hyperbolic 3-manifolds, 2008
- 908 **J.-L. Waldspurger,** L'endoscopie tordue n'est pas si tordue, 2008
- 907 **Yuanhua Wang and Fei Xu,** Spinor genera in characteristic 2, 2008
- 906 **Raphaël S. Ponge,** Heisenberg calculus and spectral theory of hypoelliptic operators on Heisenberg manifolds, 2008
- 905 **Dominic Verity,** Complicial sets characterising the simplicial nerves of strict ω-categories, 2008
- 904 **William M. Goldman and Eugene Z. Xia,** Rank one Higgs bundles and representations of fundamental groups of Riemann surfaces, 2008
- 903 **Gail Letzter,** Invariant differential operators for quantum symmetric spaces, 2008
- 902 **Bertrand Toën and Gabriele Vezzosi,** Homotopical algebraic geometry II: Geometric stacks and applications, 2008
- 901 **Ron Donagi and Tony Pantev (with an appendix by Dmitry Arinkin),** Torus fibrations, gerbes, and duality, 2008
- 900 **Wolfgang Bertram,** Differential geometry, Lie groups and symmetric spaces over general base fields and rings, 2008
- 899 **Piotr Hajłasz, Tadeusz Iwaniec, Jan Malý, and Jani Onninen,** Weakly differentiable mappings between manifolds, 2008
- 898 **John Rognes,** Galois extensions of structured ring spectra/Stably dualizable groups, 2008
- 897 **Michael I. Ganzburg,** Limit theorems of polynomial approximation with exponential weights, 2008

For a complete list of titles in this series, visit the
AMS Bookstore at **www.ams.org/bookstore/**.